SpringerBriefs in Electrical
and Computer Engineering

More information about this series at http://www.springer.com/series/10059

Yanjiao Chen • Qian Zhang

Dynamic Spectrum Auction
in Wireless Communication

Springer

Yanjiao Chen
Department of Computer Science
and Engineering
Hong Kong University of Science
and Technology
Kowloon
Hong Kong SAR

Qian Zhang
Department of Computer Science
and Engineering
Hong Kong University of Science
and Technology
Kowloon
Hong Kong SAR

ISSN 2191-8112 ISSN 2191-8120 (electronic)
SpringerBriefs in Electrical and Computer Engineering
ISBN 978-3-319-14029-2 ISBN 978-3-319-14030-8 (eBook)
DOI 10.1007/978-3-319-14030-8

Library of Congress Control Number: 2015930014

Springer Cham Heidelberg New York Dordrecht London

Printed on acid-free paper

Springer is part of Springer Science+Business Media (www.springer.com)

Preface

Static spectrum allocation hinders the efficient use of spectrum, one of the most valuable and fundamental resources for wireless communication. Spectrum auction, which enables new users to gain spectrum access and existing spectrum owners to obtain financial benefits, can greatly improve spectrum efficiency by resolving the problem of artificial spectrum shortage. However, spectrum auction design faces significant challenges due to the nature of the spectrum, including reusability, spatial and temporal availability. This book focuses on the state-of-art research on spectrum auction design, including fundamental auction theory, characteristics of spectrum market, spectrum auction architecture and possible auction mechanisms.

Hong Kong   Yanjiao Chen
August 2014   Qian Zhang

Contents

Acronyms

dB	Decibel
FCC	Federal Communications Commission
GHz	Gigahertz
IID	Independent and Identically Distributed
ITU	International Telecommunication Union
MHz	Megahertz
PCAST	President's Council of Advisors on Science and Technology
SMRA	Simultaneous Multiple Round Auction
VCG	Vickrey-Clarke-Groves

Chapter 1
Introduction

Spectrums are indispensable resources for wireless communication [34]. Propelled by the rapid development of smart devices and 4G technology, the demand for wireless traffic increases exponentially. In 2010, users worldwide downloaded 5 billion mobile applications, 15 times more than the figure (300 million) in 2009. In the U.S., the number of subscribers to mobile services increased by 20 million in 2011 alone, amounting to 294 million [3]. Such a demand will surpass the capacity of allocated wireless spectrums for mobile broadband services by as soon as 2013 [55]. To deal with this problem, on the one hand, the regulators are releasing more spectrums for commercial use; on the other hand, secondary spectrum markets emerge where incumbent spectrum licensees lease their spectrums to other service providers. In 2010, the Federal Communications Commission (FCC) in the U.S. decided to make 500 MHz of new wireless spectrum available within ten years [54]. In July 2012, the President's Council of Advisors on Science and Technology (PCAST) of the U.S. further proposed to identify 1000 MHz of federal spectrum for commercial use [51]. In 2010, the FCC introduced the idea of incentive auction to encourage incumbent spectrum licensees to voluntarily give up their license and get part of the revenue from re-selling their spectrums [3]. Company Spectrum® Bridge has launched an online platform called SpecEx for spectrum owners to sell their unused spectrums to potential buyers [1]. Spectrum auction can be an efficient way to reallocate these spectrums, either from the regulators to the wireless service providers or from incumbent spectrum licensees to secondary service providers [17].

Spectrum auction is different from traditional auction mainly due to the nature of spectrums, especially the reusability characteristic [28, 61]. A spectrum can be reused by multiple buyers if they don't interfere with each other[1] [30, 38]. Because of path loss, the transmission range of a signal is limited [4, 20]. If buyer A is beyond the transmission range of buyer B, then buyer B's transmission will not affect buyer A. The transmission range of a spectrum depends on its central frequency. By

---

[1] We assume that the entire available spectrum band are divided into spectrums with equal bandwidth. Therefore, we refer to "spectrum" as countable commodities.

© The Author(s) 2015
Y. Chen, Q. Zhang, *Dynamic Spectrum Auction in Wireless Communication*,
SpringerBriefs in Electrical and Computer Engineering, DOI 10.1007/978-3-319-14030-8_1

leveraging reusability, a spectrum can be auctioned to multiple buyers, as long as interference constraints are obeyed. This can greatly improve spectrum utilization, but poses challenges for auction design. One of the fundamental requirements for auction design is truthfulness, which means that any buyer or seller will bid their true valuations for the auctioned commodities [41, 47]. However, traditional auction mechanisms, when applied directly to spectrum auction, will become untruthful [71, 72]. In other words, auction participants have opportunities to manipulate their bids to gain higher utilities, which disrupts the economic robustness of the auction. Therefore, new auction mechanisms are need to address the spectrum reusability while maintaining nice economic properties.

Apart from spectrum reusability and economic properties, there are four other concerns in the spectrum auction design.

- *Auction Format.* Forward auction, reverse auction or double auction.
- *Demand/supply restrict.* Single item auction or multiple item auction.
- *Spectrum attribute.* Homogeneous spectrums or heterogeneous spectrums.
- *Auction dynamics.* Static auction or dynamic auction (also known as online auction).

In the forward auction, there is one seller and multiple buyers; in the reverse auction, there is one buyer and multiple sellers; in the double auction, there are multiple sellers, multiple buyers and one auctioneer. The auctioneer takes the responsibility of collecting asks from the sellers and bids from the buyers, deciding the spectrum allocation and the prices. Forward auction and double auction are the most common spectrum auction formats while reverse auction is seldom used because in common cases, there are more spectrum demands than spectrum supplies. In single item auction, each seller or buyer is restricted to trade one spectrum; while in multiple item auction, each seller or buyer is allowed to trade multiple spectrums. Multiple item auction is more flexible than single item auction, but more difficult to ensure truthfulness.

If spectrums are treated as homogeneous, there is no distinction between spectrums with different central frequencies. If spectrum heterogeneity is considered, several issues will arise. First, buyers and sellers may have different valuations for different spectrums. A spectrum with long transmission range may be suitable for large cell size (e.g. macrocell network); while a spectrum with short transmission range may be desirable for small cell size (e.g. femtocell network). The interference relationship between buyers will become quite complicated. If a buyer's device operates on a high frequency spectrum, he will interfere with a shorter range of other buyers; if a buyer's device operates on a low frequency spectrum, he will interfere with a wider range of other buyers. To decide which buyers can reuse the same spectrum becomes challenging.

In the static spectrum auction, the auction only lasts for one time stage. Static spectrum auction is suitable for long-term spectrum allocation, where the spectrum availability, the wireless environment and the interference relationship are relatively stable. In the dynamic spectrum auction, the auction will be performed for finite or infinite time stages. Dynamic spectrum auction is quite different from static spectrum auction. In the dynamic spectrum auction, the buyers may come sporadically, and the auction results in the earlier time stages will affect those in the latter time stages. For example, if a spectrum is allocated to buyer A for 2 time slots in the first stage, it cannot be allocated to other buyers who interfere with buyer A in the second stage. This makes it difficult to decide how to allocate spectrums in every time stage. To solve this problem, it is needed to estimate the influence of current spectrum allocation on the spectrum allocation in the following time stages.

In this book, we mainly focus on sealed-bid, collusion-free auction. Sealed-bid means that all bidders simultaneously submit their bids, so that no bidder knows the bids of any other bidders. Collusion-free means that no bidders collude with each other to improve the utility of the collusion group. In Chap. 5, we will discuss the problem of collusion in the spectrum auction as a future research direction. In the rest of this chapter, we will describe the background of spectrum auction in more details. In Chap. 2, we will introduce static spectrum auction mechanisms which treat spectrums as homogeneous commodities, in both forward and double auction formats. In Chap. 3, we will consider spectrum heterogeneity and introduce a static heterogeneous spectrum double auction mechanism. In Chap. 4, we will focus on online spectrum auction and introduce a dynamic heterogeneous spectrum double auction mechanism. Finally, in Chap. 5, we will give future research directions on spectrum auction.

1.1 Property of Spectrums

In this section, we will show the basic transmission model, based on which the spectrum reusability is determined.

1.1.1 Transmission Range and Spectrum Reusability

The transmission range of a spectrum determines the interference relationship among buyers, which is important for determining spectrum reusability. The power of an electromagnetic wave will decrease as it propagates through free space. The reduction of the power is usually referred to as path loss. Path loss is influenced by the environment (urban or rural), propagation medium (humidity of the air), the distance between the transmitter and the receiver, the location of the antenna, and the central frequency of the spectrum. According to the propagation model recommended by the International Telecommunication Union (ITU) [52], the path loss is affected by the central frequency of a spectrum according to the following function.

$$L = 10 \log f^2 + \gamma \log d + P_f(n) - 28 \tag{1.1}$$

in which L is the total path loss in decibel (dB), f is the central frequency of the spectrum in megahertz (MHz), d is the transmission distance in meter (m), γ is the distance power loss coefficient and $P_f(n)$ is the floor loss penetration factor. Let P_t and P_r denote the transmission power and targeted receiving power, respectively. The maximum allowable path loss is $L_{max} = P_t - P_r$. Therefore, the maximum transmission range is:

$$R_{max} = \exp\left\{\frac{P_t - P_r + 28 - P_f(n) - 10\log f^2}{\gamma}\right\} \tag{1.2}$$

Given the central frequency of a spectrum, its transmission range R_{max} can be computed by (1.2). It is obvious that, a high frequency spectrum with a larger f has a shorter transmission range, while a low frequency spectrum with a smaller f has a longer transmission range.

Assume that a transmitter operates on a spectrum with central frequency f and transmission range R_{max} determined by (1.2), other user devices within the range of R_{max} will be interfered. The interference relationship between two users is often not symmetric, even if they operate on the same spectrum [50, 59]. This is because the channel conditions between the two users is often asymmetric, i.e., the $P_f(n)$ and γ are different in (1.2).

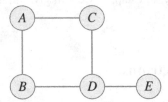

Fig. 1.1. Illustration of the interference graph

1.1.2 *Interference Graph*

Interference graph is the most common method to represent interference relationship among buyers[2]. It is an undirected graph constructed based on the transmission range of the spectrum and geographic information of the buyers [11, 56]. Therefore, interference graph is spectrum-specific. In other words, different spectrums with different central frequencies should have different interference graphs since their transmission ranges are different. Interference graph makes it easy to apply graph theory to solve the problem of spectrum reusability. Let $G = (V, E)$ denote an interference graph based on a specific spectrum. V is the set of nodes, and E is the set of edges. Each node represents a buyer. If two buyers interfere with each other, there is an edge between them; otherwise, there is no edge between them. Since

---

[2] Some works also used interference temperature instead of interference graph [24, 68].

the interference graph is undirected, it is implicitly assumed that the interference relationship between any two buyers is symmetric. Two nodes without an edge between them can reuse the same spectrum. For example, nodes A and D in Fig. 1.1. Furthermore, a group of nodes that share no edges can reuse the same spectrum. For example, nodes B, C and E in Fig. 1.1. To find such group of nodes is equivalent to finding an independent set on the interference graph, a classic problem in graph theory with many ready-to-use algorithms [6, 10, 44].

1.2 Traditional Auction Mechanisms

In this section, we briefly introduce three well-known truthful auction mechanisms. The major drawback of these auction mechanisms is that they don't consider spectrum reusability.

1.2.1 Secondary Auction

We take multiple item forward auction as an example. Secondary auction mechanism processes as follows. First, sort the buyers' bids in non-ascending order. If there are M items, name the top M buyers as winners and charge them the $(M + 1)$th buyer's bid. A simple extension of secondary auction to forward spectrum auction is shown to be untruthful in [71].

1.2.2 Vickrey-Clarke-Groves Auction

Vickrey-Clarke-Groves (VCG) auction mechanism tries to maximize social welfare with feasible allocation [16, 29, 62]. Feasible allocation refers to the auction results that satisfy the constraints of the auction (e.g., total number of auctioned items). We take forward auction as an example. Social welfare is defined as the total valuation of all the winning buyers [48]. We presume that the auction is truthful, so that the total valuation equals the total bid of all the winning buyers. Let b_i denote the bid of buyer i. First, find one optimal feasible allocation A^* that maximizes the total bid of all winning buyers (usually through brute force). For a winning buyer i, assume that in the optimal allocation, all the other buyers gain utility $\sum_{j \neq i} b_j(A^*)$. Having removed buyer i, we can find another optimal feasible allocation $\widetilde{A}^*$, all the buyers except i will gain utility $\sum_{j \neq i} b_j(\widetilde{A}^*)$. Then buyer i will be charged the price $\sum_{j \neq i} b_j(\widetilde{A}^*) - \sum_{j \neq i} b_j(A^*)$. Although VCG mechanism possesses many good properties such as truthfulness, the computational complexity is its major drawback. Approximate-VCG mechanisms have been explored [40, 42] to achieve polynomial time complexity while maintain truthfulness or approximate truthfulness. VCG auction mechanism can be proved to be truthful for traditional auction. However, a simple extension of VCG auction mechanism to forward spectrum auction is shown to be untruthful in [71], and a simple extension of VCG to double spectrum auction is shown to violate the economic property of budget balance in [72].

1.2.3   McAfee Auction

Single-item double auction [7, 18] and multi-item double auction [8, 32] mechanisms have been developed, mainly following the idea of McAfee [45]. Assume that each buyer or seller has one item to trade. First, sort the sellers' asks in non-descending order and sort the buyers' bids in non-ascending order (sellers' bidding prices are often referred to as "asks"; while buyers' bidding prices are often referred to as "bids"). Then, find index k so that the kth seller's ask is no greater than the kth buyer's bid, but the $(k + 1)$th seller's ask is strictly greater than the $(k + 1)$th buyer's bid. After doing so, the first $(k - 1)$ buyers and $(k - 1)$ sellers become winners. Each winning seller is paid by the kth seller's ask; and each winning buyer pays by the kth buyer's bid. The static homogeneous spectrum double auction we introduce in Chapter 2 follows the design rationale of McAfee, but carefully design the spectrum allocation and pricing mechanisms to enable spectrum reusability and guarantee truthfulness.

1.3   Economic Properties

In this section, we introduce three economic properties that are deemed to be most essential for spectrum auction design.

1.3.1   Truthfulness

Truthfulness is one of the most fundamental property of an auction mechanism [39]. The buyers and sellers are selfish and rational players, who will manipulate their asks and bids to maximize their own utilities. Being truthful means that a seller's ask or a buyer's bid equal their true valuations for the spectrum[3]. A truthful auction mechanism guarantees that a buyer or a seller cannot get higher payoff by misreporting their true valuations, thus they will have no incentive to be untruthful. For the online spectrum auction, we have to further consider truthfulness at each time stage.

---

[3] A broader meaning of truthfulness may also include that a buyer truthfully reports his spectrum demand or time slot requirement.

1.3.2 Individual Rationality

A buyer or a seller is individually rational in the sense that they will not participate in the auction, if by doing so their utilities become negative. An auction mechanism is individually rational, if all sellers and buyers achieve non-negative utility. In other words, in an individual rational auction, any seller is paid more than his ask, and any buyer pays less than his bid.

1.3.3 Budget Balance

Budget balance is often considered in the double auction. It means that the auctioneer maintains non-negative budget. In other words, the money that the auctioneer gets from all buyers is no less than the money he gives to all sellers. For regulators, budget balance is often enough to motivate them to host spectrum auctions. The profit-oriented auctioneers, however, may aim at revenue maximization.

Ideally, we want the auction mechanisms to possess the above three economic properties while maximizing spectrum utilization via spectrum reuse. However, it has been proved that no double auction mechanism can achieve highest spectrum utilization and maintain economic properties at the same time [49, 69]. Most of the auction mechanisms target at economic robustness [7, 8, 18, 32, 72]. Some spectrum auction mechanisms also aim at revenue maximization [5, 25, 53] or collusion resistance [66–68].

Chapter 2
Static Homogeneous Spectrum Auction

In this chapter, we will introduce static homogeneous spectrum auction, first in forward auction format, then in double auction format. The key of the auction design is to leverage spectrum reusability while guaranteeing economic properties.

2.1 Homogeneous Spectrum Forward Auction

In the homogeneous spectrum forward auction, we assume that there is one seller who owns k spectrums, and there are N buyers. Buyer i's bid consists of two parts: d_i is the number of spectrums wanted and b_i is the bidding price. In homogeneous spectrum auction, buyer i's bidding price is the same for all spectrums. For strict request, buyer i either accepts d_i spectrums or 0 spectrum; for range request, buyer i accepts any spectrums in the range $[0, d_i]$. Here, we consider strict request. The bid of a buyer is based on his true valuation for the spectrum, denoted by v_i. In the homogeneous spectrum auction, buyer i has the same valuation for all spectrums. If buyer i becomes a winner, the seller will charge him p_i for each spectrum; otherwise, the seller will charge him nothing. Buyer i's utility is his valuation for the obtained spectrum minus the paid price.

$$U_i = \begin{cases} (p_i - v_i) * d_i, & \text{if buyer } i \text{ wins} \\ 0, & \text{otherwise} \end{cases} \tag{2.1}$$

2.1.1 A Naive Truthful Auction Mechanism

We first introduce a naive auction mechanism based on secondary auction, which is truthful but greatly reduces spectrum utilization. As shown in Fig. 2.1, suppose that all the buyers are located in a rectangular region. The maximum transmission range of all the spectrums is R_{max}. We partition the whole region into small squares

© The Author(s) 2015
Y. Chen, Q. Zhang, *Dynamic Spectrum Auction in Wireless Communication*,
SpringerBriefs in Electrical and Computer Engineering, DOI 10.1007/978-3-319-14030-8_2

of $R_{max} \times R_{max}$ area. Therefore, buyers in two non-adjacent squares do not interfere with each other. We divide the spectrums into four sets, each set containing $k/4$ spectrums. In each square, we apply secondary auction, allocating $k/4$ spectrums to the top $k/4$ bidders within the square and charge them the price of the $(k/4 + 1)$th highest bid. Every two adjacent squares will have different sets of spectrums. In this way, the spectrum allocation will be interference free, and the spectrums can be partially reused. However, not every two buyers in a square interfere with each other. Since only $1/4$ of the total spectrums are allocated in each square, the spectrum utilization will be low.

2.1.2 Auction Mechanism Based on Greedy Algorithm

To improve spectrum utilization, an auction mechanism based on greedy algorithm VERITAS is proposed in [71]. The major algorithm includes two parts: spectrum allocation and price determination.

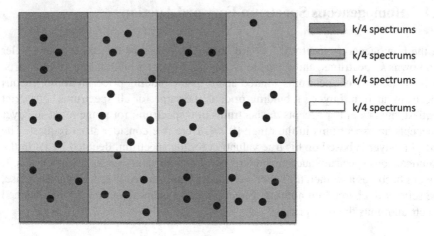

Fig. 2.1. A naive truthful auction mechanism

Spectrum Allocation Algorithm 1 shows the procedure of spectrum allocation. To begin with, the buyers are sorted according to their bids in non-ascending order. In each iteration, the unassigned buyer with the highest bid is considered. Let $N(i)$ denote the set of interfering neighbors of buyer i. If his spectrum demand d_i is fewer than the number of available spectrums, that is, k minus the spectrums assigned to his interfering neighbors $\sum_{j \in N(i)} d_j$, his demand can be fulfilled; otherwise, he will not be assigned any spectrums because of strict request.

Price Determination Algorithm 2 is the procedure of price determination. The idea of price determination is to charge a buyer the unit price which equals the bid of his critical neighbor.

Definition 1 *Critical neighbor.* The critical neighbor of buyer i is one of his interfering neighbors $j \in N(i)$. If buyer i bids no smaller than buyer j, that is, $b_i \geq b_j$, buyer i will be allocated d_i spectrums; otherwise, buyer i will be allocated no spectrums.

Algorithm 1 Spectrum allocation in VERITAS

1: $B =$ Sorted list of buyers according to their bids in non-ascending order.
2: **while** $B \neq \Phi$ **do**
3:　　i is the first buyer in B.
4:　　**if** $d_i \leq M - \sum_{j \in N(i)} d_j$ **then**
5:　　　　Assign d_i spectrums to buyer i.
6:　　**end if**
7:　　Remove buyer i from B.
8: **end while**

Algorithm 2 shows how to find the critical neighbor for a buyer in an efficient way. To calculate the price for buyer i, firstly, buyer i is removed from the sorted buyer list B. Then, the algorithm runs like Algorithm 1, allocating spectrums to buyers iteratively. Every time an interfering neighbor j of buyer i is allocated spectrums, we check whether the rest of the spectrums is sufficient for buyer i. If not, buyer j is the critical neighbor of buyer i. This is because if buyer i bids lower than buyer j, he will be placed behind buyer j in the sorted list B. When it comes to buyer i's iteration, there will not be enough spectrums for him, and he will be allocated no spectrums.

Algorithm 2 Price determination for buyer i in VERITAS

1: $p_i = 0$.
2: **if** $d_i = 0$ **then**
3:　　Return.
4: **end if**
5: $B' = B \setminus \{b_i\}$.
6: $Avail = k$.
7: **while** $B' \neq \Phi$ AND $flag = 1$ **do**
8:　　j is the first buyer in B'.
9:　　**if** $d_j \leq M - \sum_{l \in N(j)} d_l$ **then**
10:　　　　Assign d_j spectrums to buyer j.
11:　　　　**if** $j \in N(i)$ **then**
12:　　　　　　$Avail = Avail - d_j$.
13:　　　　　　**if** $Avail < d_i$ **then**
14:　　　　　　　　$p_i = b_j$.
15:　　　　　　　　$flag = 0$.
16:　　　　　　**end if**
17:　　　　**end if**
18:　　**end if**
19:　　Remove buyer j from B'.
20: **end while**

2.1.3   Proofs of Economic Properties

Individual Rationality   To prove that the buyers have non-negative utility, we only have to prove the following proposition.

Proposition 1   *The unit price charged from a buyer is always smaller than his bid.*

Proof   If $p_i = 0$, it is clear that $p_i \leq b_i$. If $p_i > 0$, it means that buyer i is allocated spectrums and p_i is the bid of his critical bidder. According to Definition 1 of the critical bidder, $b_i \geq p_i$, otherwise buyer i cannot win any spectrums. Therefore, we have proved that the unit price charged from a buyer is always smaller than his bid.

Truthfulness   To prove the truthfulness of the auction, we only have to prove the following proposition.

Table 2.1   Possible auction results

Case	I	II	III	IV
The seller/buyer is truthful	Lose	Win	Win	Lose
The seller/buyer is untruthful	Lose	Lose	Win	Win

Proposition 2   *Any buyer cannot gain higher utility when he bids untruthfully than when he bids truthfully.*

Proof   When a buyer i bids truthfully and untruthfully, possible auction results are listed in Table 2.1. Let U_i and p_i denote buyer i's utility and price when he bids truthfully; U_i' and p_i' denote buyer i's utility and price when he bids untruthfully; v_i denote buyer i's true valuation for each spectrum. We prove that in every case, $U_i \geq U_i'$.

- Case I. As buyer i loses when he bids truthfully and untruthfully, $U_i = U_i' = 0$.
- Case II. As buyer i loses when he bids untruthfully, $U_i' = 0$. As buyer i wins when he bids truthfully, $U_i \geq 0$ according to individual rationality. Therefore, $U_i \geq U_i'$.
- Case III. As buyer i wins when he bids truthfully and untruthfully, p_i and p_i' both equal the bid of his critical neighbor, therefore $U_i = U_i'$.
- Case IV. As buyer i loses when he bids truthfully, $U_i = 0$. Also, this means that his bid, which equals his true valuation v_i, is smaller than the bid of his critical neighbor. When buyer i wins by bidding untruthfully, it must be true that $p_i' > v_i$ because p_i' equals the bid of his critical neighbor. Therefore we have $U_i' = (v_i - p_i') * d_i < 0 = U_i$.

In summary, we have proved that $U_i \geq U_i'$ for all possible auction results. Therefore, a buyer has no incentive to bid untruthfully in the auction.

2.2 Homogeneous Spectrum Double Auction

In the homogeneous spectrum double auction, we assume that there are M sellers and N buyers, and that each seller owns one spectrum and each buyer wants to buy one spectrum. We assume that the spectrums are available to all buyers. Let b_i^s, v_i^s, U_i^s denote the ask, true valuation and utility of seller i; b_j^b, v_j^b, U_j^b denote the bid, true valuation and utility of buyer j; U^a denote the utility of the auctioneer. After the auction, the auctioneer pays seller i price p_i^s and charges buyer j price p_j^b. The seller i's utility is his payment minus his true valuation for the spectrum.

$$U_i^s = \begin{cases} p_i^s - v_i^s, & \text{if seller } i \text{ wins} \\ 0, & \text{otherwise} \end{cases} \tag{2.2}$$

The buyer j's utility is his true valuation for the spectrum minus his price.

$$U_j^b = \begin{cases} v_j^b - p_j^b, & \text{if buyer } j \text{ wins} \\ 0, & \text{otherwise.} \end{cases} \tag{2.3}$$

The auctioneer's utility is his collected payment from all buyers minus his payment to all sellers.

$$U^a = \sum_j p_j^b - \sum_i p_i^s. \tag{2.4}$$

2.2.1 Auction Mechanism Design

In this section, we introduce a truthful spectrum double auction mechanism TRUST [72, 73], which consists of three processes: grouping, spectrum allocation, and price determination.

Grouping The auctioneer groups the buyers who can reuse the same spectrum together. To find such buyer groups, the auctioneer can construct interference graph of all buyers, then find independent sets on the graph. To avoid market manipulation, the grouping process is performed privately by the auctioneer, unknown by neither the sellers or the buyers. Assume there are L buyer groups. Let $g_1, g_2, ..., g_L$ denote the resulting groups, and $|g_i|$ is the number of buyers in group g_i. If g_i is allocated a spectrum, all the members in g_i will reuse the spectrum.

Spectrum Allocation Algorithm 3 shows the process of spectrum allocation. First, the bid of a group is calculated based on the lowest bid in the group. Then, the sellers are sorted according to their asks in non-descending order, and the buyer groups are sorted according to their group bids in non-ascending order. The last profitable trade is the kth one in which the kth buyer group's bid is no less than the kth seller's ask, but the $(k + 1)$th buyer group's bid is strictly smaller than the $(k + 1)$th seller's ask. The first $(k - 1)$ sellers and $(k - 1)$ buyer groups are the winners. Since the spectrums are homogeneous, we can randomly assign the $(k - 1)$ spectrums to the $(k - 1)$ winning groups.

Algorithm 3 spectrum allocation in TRUST

1: Calculate the group bid of g_i as

$$B_i = |g_i| \times \min_{j \in g_i} b_j^b \qquad (2.5)$$

2: Sort the sellers according to their asks in non-descending order.
3: Sort the buyer groups according to their bids in non-ascending order.
4: Find k that $B_k \geq b_k^s$ and $B_{k+1} < b_{k+1}^s$.
5: The auction winners are the top $k - 1$ sellers and the buyers in the top $k - 1$ groups.
6: Randomly allocate the spectrums of the top $k - 1$ sellers to the top $k - 1$ buyer groups.

Price Determination If a seller or a buyer is not a winner, he will be paid or pay nothing. If seller i is among the top $k - 1$ sellers, the auctioneer will pay seller i the price of the kth seller.

$$p_i^s = b_k^s. \qquad (2.6)$$

If buyer group g_j is among the top $k - 1$ buyer groups, the auctioneer will charge buyer group g_j the price of the kth buyer group, which will shared by all members in the group.

$$p_l^b = B_k / |g_j|, \forall l \in g_j. \qquad (2.7)$$

2.2.2   Proofs of Economic Properties

Individual Rationality

Proposition 3 *The auction mechanism TRUST is individual rational for both buyers and sellers.*

Proof If a buyer or a seller is not a winner, his utility is zero (non-negative).

If seller i is a winner, he is paid $p_i^s = b_k^s \geq b_i^s$ since the sellers are sorted in non-descending order.

If buyer l is a winner in group g_i, he pays $p_l^b = B_k / |g_i| \leq B_i / |g_i| = \min_{j \in g_i} b_j^b \leq b_l^b$.

Budget Balance

Proposition 4 *The auctioneer's utility is non-negative.*

Proof The auctioneer's utility is as follows.

$$U^a = (k - 1)(B_k - b_k^s) \geq 0$$

Truthfulness We have to prove truthfulness both on buyers' side and on sellers' side. According to Table 2.1, there are four possible auction results.

Proposition 5 *The auction mechanism is truthful for sellers.*

Proof Let U_i^s and p_i^s denote seller i's utility and price when he bids truthfully; and $U_i^{s'}$ and $p_i^{s'}$ denote seller i's utility and price when he bids untruthfully; v_i^s denote seller i's true valuation for the spectrum. We prove that in every case, $U_i^s \geq U_i^{s'}$.

- Case I. As seller i loses when he bids truthfully and untruthfully, $U_i^s = U_i^{s'} = 0$.
- Case II. As seller i loses when he bids untruthfully, $U_i^{s'} = 0$. As seller i wins when he bids truthfully, $U_i^s \geq 0$ according to individual rationality. Therefore, $U_i^s \geq U_i^{s'}$.
- Case III. We first prove that if seller i wins when he bids truthfully and untruthfully, his payments p_i^s and $p_i^{s'}$ are the same.
  When seller i bids truthfully, the sorted seller list is $\{b_1^s, ..., b_k^s, b_{k+1}^s, ...\}$, and the sorted buyer group list is $\{B_1, ..., B_k, B_{k+1}, ...\}$. We have $b_k^s \leq B_k$ and $b_{k+1}^s > B_{k+1}$. When seller i bids untruthfully, it can be easily proved that his bid cannot be greater than b_k^s, otherwise seller i cannot be a winner. Therefore, the top $k - 1$ seller and the kth seller do not change. So we have $p_i^s = p_i^{s'} = b_k^s$, and $U_i^s = p_i^s - v_i^s = p_i^{s'} - v_i^s = U_i^{s'}$.
- Case IV. As seller i loses when he bids truthfully, $U_i^s = 0$. The sorted seller list is $\{b_1^s, ..., b_k^s, b_{k+1}^s, ..., b_i^s, ...\}$, and the sorted buyer group list is $\{B_1, ..., B_k, B_{k+1}, ...\}$. We have $b_k^s \leq B_k$ and $b_{k+1}^s > B_{k+1}$. Since seller i loses, we know that $b_i^s \geq b_k^s$. If seller i wins by bidding untruthfully, it must be true that $b_i^{s'} < b_k^s$. The sorted seller list becomes $\{b_1^s, ..., b_i^{s'}, ..., b_{k-1}^s, b_k^s, ...\}$, and the sorted buyer group list does not change. B_k and b_{k-1} become a pair. As $b_{k-1}^s \leq b_k^s \leq B_k$, the final payment for all sellers will be no greater than b_{k-1}^s. Therefore, $p_i^{s'} \leq b_{k-1}^s \leq b_k^s \leq b_i^s = v_i^s$. Hence, $U_i^{s'} = p_i^{s'} - v_i^s \leq 0 = U_i^s$.

Proposition 6 *The auction mechanism is truthful for buyers.*

Proof Let U_i^b and p_i^b denote buyer i's utility and price when he bids truthfully; and $U_i^{b'}$ and $p_i^{b'}$ denote buyer i's utility and price when he bids untruthfully. We prove that in every case, $U_i^b \geq U_i^{b'}$.

- Case I. Similar to seller's CASE I.
- Case II. Similar to seller's CASE II.
- Case III. We first prove that if buyer i wins when he bids truthfully and untruthfully, his price p_i^b and $p_i^{b'}$ are the same.
  When buyer i bids truthfully, the sorted seller list is $\{b_1^s, ..., b_k^s, b_{k+1}^s, ...\}$, and the sorted buyer group list is $\{B_1, ..., B_k, B_{k+1}, ...\}$. We have $b_k^s \leq B_k$ and $b_{k+1}^s > B_{k+1}$.

When seller i bids untruthfully, it can be easily proved that his group's bid cannot be smaller than B_k, otherwise, his group cannot be a winner. Therefore, the top $k - 1$ buyer groups and the kth buyer group do not change. So we have $p_i^b = p_i^{b'} = B_k/|g_j|, i \in g_j$, and $U_i^b = p_i^b - v_i^b = p_i^{b'} - v_i^b = U_i^{b'}$.

- Case IV. When buyer i bids truthfully, the sorted seller list is $\{b_1^s, ..., b_k^s, b_{k+1}^s, ...\}$, and the sorted buyer group list is $\{B_1, ..., B_k, B_{k+1}, ..., B_j, ...\}$. Since buyer i loses, we know that $B_j \leq B_k$. We have $U_i^b = 0$. As buyer i loses by bidding truthfully and wins by bidding untruthfully, it must be true that b_i^b is the minimum in group g_j, and $b_i^{b'} > b_i^b$ to change the group bid so that $B_j' > B_k$. When buyer i bids untruthfully, the sorted seller list does not change, and the sorted buyer group list becomes $\{B_1, ..., B_j', ..., B_{k-1}, B_k, ...\}$. B_{k-1} and b_k become a pair. As $B_{k-1} \geq B_k \geq b_k^s$, the final price for all buyer groups will be no less than B_{k-1}. Therefore, $p_i^{b'} \geq B_{k-1}/|g_j| \geq B_j/|g_j| \geq b_i^b = v_i^b$. Hence, $U_i^{b'} = v_i^b - p_i^{b'} \leq 0 = U_i^b$.

Chapter 3
Static Heterogeneous Spectrum Auction

Existing works on double spectrum auction mostly treat spectrums as identical commodities with homogeneous attributes [13, 72], which is acceptable when a small number of spectrums are considered. However, with the forthcoming freeing-up of large pool of spectrums, especially the ones used in TV service that have wide range frequencies, the homogeneous spectrum auction will suffer from severe utilization loss and interference problems.

Spectrums are intrinsically heterogeneous because signal propagation is frequency-selective. As we discussed in Sect. 1.1.1, Chap. 1, spectrums of different frequencies have different path losses, thus different transmission ranges. Heterogeneous spectrum auction is challenging due to the following reasons.

- Interference relationship among buyers will be different when different spectrums are used for transmission. Buyers who are interference-free on a high frequency spectrum may generate harmful interference to each other on a low frequency spectrum.
- The co-existence of macro/micro/pico/femto cellular networks indicates that network operators have different targeted cell coverages, therefore different preferences for different frequencies. Macrocell operators may favor a low frequency spectrum with a long transmission range while femtocell operators are satisfied with a high frequency spectrum whose transmission range is enough to cover the indoor area and generates less cross-tier interference.

A naive way to deal with spectrum heterogeneity is to conduct a separate auction for each heterogeneous spectrum. However, this is not only inefficient (e.g., 10 auctions for 10 heterogeneous spectrum), but also makes it hard for a buyer to control how many spectrums he wants and how many spectrums he can win. For example, a buyer who wants to buy one spectrum submits different bids to different auctions. If his bids are the highest in multiple auctions, he may win more spectrums than he wishes for. Therefore, we need an integrated auction where the buyers can bid for all heterogeneous spectrums simultaneously.

Unfortunately, spectrum heterogeneity makes it difficult to design an economic-robust auction mechanism that achieves high spectrum utilization. If bidders are forced to bid a common price for all spectrums, they may overpay for a less-valued

Y. Chen, Q. Zhang, *Dynamic Spectrum Auction in Wireless Communication*, SpringerBriefs in Electrical and Computer Engineering, DOI 10.1007/978-3-319-14030-8_3

17

spectrum when bidding high, violating the property of individual rationality; they may lose a high-valued spectrum when bidding low, not reflecting their true valuation for the spectrum. If bidders are granted the right to bid different prices for different spectrums, they will have the opportunity to manipulate the price for one spectrum to influence the winning results of another spectrum, which violates the property of truthfulness. In addition, interference graph heterogeneity makes it hard to exploit spectrum reusability while ensuring interference-free spectrum allocation.

Existing spectrum auction mechanisms, though having resolved the problem of spectrum reusability, failed to consider spectrum heterogeneity. Most of existing works treated spectrums as identical objects [5, 35, 63, 71, 72, 74] and assumed that a buyer submits a uniform price for all spectrums. In this way, buyers either fail to bid higher and have higher chance to win a higher-valued spectrum or become unsatisfactory for obtaining a lower-valued spectrum. Due to a lack of flexible bidding, the sellers' revenue is also affected. When determining whether two buyers can reuse the same spectrum, previous works either assumed that the interference graph is a complete graph [63] or used the same interference graph [5, 35, 53, 71, 72] for all spectrums. In the former case, no user can reuse the same spectrum and spectrum utilization is extremely low. In the latter case, if the interference graph is constructed based on the low frequency spectrum with long transmission range, interference constraint will be over strict, thus spectrum utilization is low; if the interference graph is constructed based on the high frequency spectrum with short transmission range, the auction results cannot guarantee interference-free spectrum allocation.

In this chapter, we will introduce two heterogenous spectrum double auction mechanisms. The first one considers the single item case, and the second one considers the multiple item case. To begin with, we describe the model of heterogenous spectrum double auction and its challenges.

3.1 Modeling Heterogeneous Spectrum Double Auction

We consider the scenario where there are N buyers and M sellers. Seller i contributes $n_i \geq 1$ spectrums. The total number of spectrums contributed by all sellers is $K = \sum_{i=1}^{M} n_i$. Let $S = (s_1, s_2, ..., s_K)$ denote the set of all the spectrums for sale. In single item heterogeneous spectrum auction, each seller contributes only one spectrum, that is, $K = M$. Seller i submits his ask to the auctioneer, denoted by $R_i = (r_{i,1}, r_{i,2}, ..., r_{i,n_i})$. Seller i's ask is based on his true valuation for the spectrum, denoted by $(v_{i,1}^s, v_{i,2}^s, ..., v_{i,n_i}^s)$.

Each buyer submits his bid to the auctioneer, denoted by $B_i = (b_{i,1}, b_{i,2}, ..., b_{i,K})$, a bidding profile in which $b_{i,j}$ is buyer i's bidding price for $s_j, j \in [1, K]$. The number of spectrums that buyer i wants is d_i. In single item heterogenous spectrum auction, $d_i = 1$; and in multiple item heterogeneous spectrum auction, $d_i \geq 1$. The buyers have different valuations for different spectrums, based on his true valuation for each spectrum, denoted by $(v_{i,1}^b, v_{i,2}^b, ..., v_{i,K}^b)$. Since buyers scatter at different areas and not all spectrums are available in each area [22], we assume that $v_{i,j}^b = 0$ if spectrum $s_j, j \in [1, K]$ is not available to buyer i. We assume that the spectrum availability to a buyer is known by the auctioneer, hence he cannot lie about it. Let $R = (R_1, R_2, \ldots, R_M)$ denote the ask matrix of all sellers and $R \backslash \{i\}$ denote the ask matrix with seller i's ask R_i removed. Let $B = (B_1, B_2, \ldots, B_N)$ denote the bid matrix of all buyers. Let $B \backslash \{i\}$ denote the bid matrix with buyer i's bid B_i removed.

We use $(w_{i,1}^s, \ldots, w_{i,n_i}^s)$ to represent seller i's winning profile, and $(p_{i,1}^s, \ldots, p_{i,n_i}^s)$ to represent seller i's payment profile. $w_{i,j}^s = 1$ means that seller i successfully sells spectrum $s_{i,j}, j \in [1, n_i]$; otherwise $w_{i,j}^s = 0$. After the auction, the utility of seller i is the payment he receives minus his true valuation for all the sold spectrums.

$$U_i^s = \sum_{j=1}^{n_i} p_{i,j}^s - \sum_{j=1}^{n_i} w_{i,j}^s * v_{i,j}^s. \tag{3.1}$$

We use $(w_{i,1}^b, \ldots, w_{i,K}^b)$ to represent buyer i's winning profile and $(p_{i,1}^b, \ldots, p_{i,K}^b)$ to represent buyer i's payment profile. $w_{i,j}^b = 1$ means that buyer i wins spectrum $s_j, j \in [1, K]$; otherwise $w_{i,j}^b = 0$. After the auction, the utility U_i^b of the buyer i is the aggregation of his true valuation for the procured spectrums minus his total payment.

$$U_i^b = \sum_{j=1}^{K} w_{i,j}^b * v_{i,j}^b - \sum_{j=1}^{K} p_{i,j}^b. \tag{3.2}$$

We consider the range request, and assume that the utility of the buyer does not depend on whether his demand is fully satisfied or not. The auction mechanisms that we introduce in the following sections guarantee that a buyer will not win more spectrums than he demands. However, he may win fewer spectrums than he demands.

3.2 Challenges of Heterogeneous Spectrum Auction Design

In this section, we demonstrate the challenges of designing a truthful heterogeneous spectrum auction mechanism. We will first introduce the heterogeneity nature of spectrums. Then we show that the buyers can manipulate their bids for higher utility in heterogeneous spectrum auction.

3.2.1 Spatial Heterogeneity

Spatial heterogeneity means that spectrum availability varies in different locations. For example, one TV spectrum is available only if there are no nearby TV stations or wireless microphones occupying the same spectrum.

Homogeneous spectrum auction addresses spectrum reusability by finding independent sets on their interference graph as buyer groups [60], and then compute the
group bid according to the minimum bid in the group [65]. However, if two buyers
with no common available spectrums are grouped together, the group bid will be 0
for all spectrums and this group can never win in the auction. For example, in Fig.
3.1, there are 6 buyers and 3 spectrums $\{a, b, c\}$. Two nodes with an edge between
them mutually interfere with each other. The bids are depicted along with each buyer.
Only spectrum b is available to buyer A, so only his bid for spectrum b is non-zero.

In this example, by finding the independent sets on the interference graph, the
6 buyers can be grouped into two groups, $\{A, D, E\}$ and $\{B, C, F\}$, respectively.
However, there is no common available spectrum for all buyers in either group. For
example, in the group $\{A, D, E\}$, spectrum b is the only common available spectrum
for the two buyers A and D, but it is not available for buyer E.

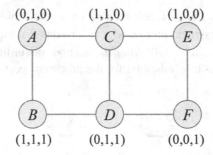

Fig. 3.1 No spectrums are commonly available for buyers in both buyer groups $\{A, D, E\}$ or
$\{B, C, F\}$

3.2.2   Frequency Heterogeneity

The distinctive property of spectrums is reusability. The auctioneer can pick multiple
non-interfering buyers for one spectrum based on the interference graph. However,
the interference graphs for spectrums with different central frequencies are no longer
the same. The spectrums offered by spectrum owners may consist of a wide range of
frequencies. For example, in the German spectrum auction held in 2010, the highest
frequency (2.6 GHz) was more than three times higher than the lowest frequency
(800 MHz) [2]. According to Eq. (1.1), this causes a more than 10 dB path loss
difference at the same distance between the two. This huge gap leads to non-identical
interference relationships among buyers on different spectrums. For example, Fig. 3.2
shows interference graphs of three buyers in terms of a high frequency spectrum a
and a low frequency spectrum b. If we use the interference graph of a and assign b
simultaneously to buyer A and C, they end up interfering with each other. If we use
the interference graph of b and assign a to only A, the spectrum utilization becomes

low. With large number of spectrums and buyers, the problem will be even more complicated.

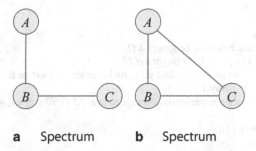

a   Spectrum   **b**   Spectrum

Fig. 3.2 Heterogeneous interference graph based on high frequency spectrum *a* and low frequency spectrum *b* respectively

3.2.3 Market Manipulation

Two well-known auction mechanisms VCG and McAfee are truthful for single item homogenous spectrum auction. However, they can be proved to be untruthful for single item heterogeneous spectrum auction. Moreover, the VCG scheme fails to achieve budget balance [70, 72].

In multiple item heterogeneous spectrum auction, a buyer may want more than one spectrums. One solution is to replace the original buyer with multiple dummies and each dummy requests one spectrum. After dummy creation, grouping can be performed in the same way as in a single item auction. However, this simple conversion leaves the chance for buyers to manipulate their bids to gain higher utility. A buyer with multiple dummies can manipulate bidding price for one spectrum to influence the winning result of another spectrum. A multiple item homogenous auction will not have this problem because a buyer bids a common price for all spectrums so that bid manipulation does not exist.

3.3 Single Item Heterogeneous Spectrum Auction

In this section, we introduce a truthful single item heterogeneous spectrum auction mechanism TAHES [21], which consists of three processes: grouping, matching, and winner and price determination.

Algorithm 4 Grouping in TAHES

1: L is the set of grouped buyers. G is the set of groups. F is the set of candidate spectrums for
   each group.
2: $L = \emptyset$, $G = \emptyset$, $F = \emptyset$.
3: **while** $|L| \neq N$ **do**
4:    **for all** $s_i \in S$ **do**
5:       Set of candidate buyers to be grouped Q.
6:       $Q = \{\text{Buyer } k | a_{k,i} = 1 \text{ AND Buyer } k \notin L\}$.
7:       Randomly choose an independent set g_i on buyer set Q based on E.
8:       $L = L \bigcup \{\text{Buyer } j | \text{Buyer } j \in g_i\}$.
9:       $f_i = \{s_j | s_j \text{ has shorter transmission range than } s_i \text{ AND } \forall \text{Buyer } k \in g_i, a_{k,j} = 1\}$.
10:   **end for**
11:   $G = G \bigcup g_i$, $F = F \bigcup f_i$;
12: **end while**
13: Return (G, F);

3.3.1 Auction Mechanism Design

To handle both spectrum heterogeneity and spectrum reusability, TAHES designs
three key steps.

Grouping Spectrums can be reused by non-interfering buyers in different locations.
By non-interfering, we mean that when two buyers operate on the same spectrum,
they are out of the transmission range of each other. However, the interference
relationship between one pair of buyers are non-identical on different frequencies.
In this step, the auctioneer uses Algorithm 4 to form non-interfering buyer groups
based on heterogeneous interference graphs so that the buyers in the same group can
purchase the same spectrum. The buyer grouping process is independent from buyers'
bids. The input of the grouping algorithm is the spectrum availability information
of each buyer. This kind of information can be calculated according to the path loss
model given the locations of buyers and sellers or can be obtained from a geo-location
database [23]. The auctioneer can get such location information of all the buyers and
sellers. The bid-independent property of the grouping algorithm is critical to ensure
truthfulness in the auction [72].

Let $A = \{a_{i,j} | a_{i,j} \in \{0, 1\}\}_{N \times M}$, an N by M matrix, represent the buyers'
spectrum availability. $a_{i,j} = 1$ means that spectrum s_j is available to buyer i. Let
$E = \{e_{i,j,k} | e_{i,j,k} \in \{0, 1\}\}_{M \times N \times N}$, an M by N by N matrix, represent the interference relationships between buyers on each spectrum. $e_{i,j,k} = 1$ means that buyers j
and k interfere on spectrum s_i.

In this step, the inputs of Algorithm 4 are A and E, which are both bid-independent.
After grouping, we get buyer groups denoted as $G = \{g_1, g_2, \ldots, g_L\}$, and the corresponding candidate spectrum set for each group denoted as $F = \{f_1, f_2, \ldots, f_l\}$. f_i
contains the spectrums that group g_i can purchase, which is assigned by the auctioneer. G and F are the outputs of Algorithm 4. The grouping process should satisfy
the following constraints:

Algorithm 5 Matching in TAHES

1: S is the set of spectrums. G is the set of groups. Δ is a $|G|$ by $|S|$ matrix representing the weighted adjacent matrix between G and S.
2: $\Delta = \{0\}_{M \times N}$;
3: **for all** $g_x \in G, s_y \in S$ **do**
4:    **if** $\delta_x^y > 0$ and $s_y \in f_x$ **then** $\Delta_{x,y} = \Delta_{y,x} = |g_x|$;
5: **end for**
6: $(G_C, S_C, \sigma) = MATCH(X, Y, \Delta)$;
7: return (G_C, S_C, σ);

- *Common spectrum existence constraint:* There exists at least one spectrum that is available for all buyers in the same group.

$$\forall g_i, \exists s_j, s.t. \forall \text{Buyer } k \in g_i \Rightarrow a_{k,j} = 1 \tag{3.3}$$

- *Interference free constraint:* Any two buyers in the same group do not mutually interfere with each other in all spectrums in the candidate spectrum set.

$$\forall g_i, \forall s_j \in f_i, \forall \text{Buyer } m, n \in g_i \Rightarrow e_{j,m,n} = 0 \tag{3.4}$$

In Algorithm 4, we can use any existing algorithms to find independent sets, for example, the algorithms described in [58]. The buyer groups and candidate spectrum sets returned by Algorithm 4 satisfy the common channel existence constraint and the interference free constraint.

Matching After the first step, each buyer group may still purchase spectrums from multiple sellers if the buyers in a group have more than one common spectrums. However, in fact, each group can win at most one spectrum. Only one bid in the bid vector should be kept non-zero for further winner determination. Otherwise, the auction can be vulnerable to market manipulation. Some buyers can strategically change some of their bids to lower the group bids so that they can change the price they need to pay and increase their utility. In this matching step, the auctioneer chooses one conventional matching algorithm to match each buyer group to only one spectrum based only on the spectrum availability information for each group. This ensures that the matching step is also bid-independent.

After the grouping process, we have formed a group set G. Let $|g_i|$ denote the number of buyers in group g_i, and $\delta_i = (\delta_{i,1}, \delta_{i,2}, \ldots, \delta_{i,M})$ denote the group bid vector. The group bid for a spectrum is the minimum bid for that spectrum times the group size.

$$\delta_{i,j} = |g_i| \times \min_{k \in g_i} b_{k,j} \tag{3.5}$$

In δ_i, there may be more than one non-zero entry. To guarantee auction truthfulness, we apply a spectrum matching scheme to match one buyer group to one seller. The results after matching are the candidate winning group set G_C and candidate winning seller set S_C. The matching procedure is shown in Algorithm 9. In this algorithm, σ records the matching result. For example, $\sigma(g_x) = s_y$ indicates that buyer group g_x is assigned to seller s_y. $MATCH(X, Y, \Delta)$ matches nodes set X to Y with weighted edges in Δ. $MATCH(X, Y, \Delta)$ can be any matching algorithm for bipartite graphs specified by the auctioneer, for example, maximum matching[19] or maximum weighted matching [46]. This matching step here is also independent of the buyers' bids and the sellers' asks.

Algorithm 6 Winner and price determination in TAHES

1: G_W and S_w are the sets of winning buyer groups and winning sellers.
2: Construct $X = \{g_{i_1}, g_{i_2}, \cdots, g_{i_L}\}$, such that $\delta_{i_1, \sigma(i_1)} \geq \delta_{i_2, \sigma(i_2)} \geq \cdots \geq \delta_{i_L, \sigma(i_L)}$.
3: Construct $Y = \{s_{j_1}, s_{j_2}, \cdots, s_{j_M}\}$, such that $R_{j_1} \leq R_{j_2} \leq \cdots \leq R_{j_M}$.
4: Find the largest k, s.t. $\delta_{i_k, \sigma(i_k)} \geq R_{j_k}$.
5: **if** $k < 2$ **then**
6:     $G_W = \emptyset$, $S_W = \emptyset$.
7:     $p_i^b = 0, p_j^s = 0, , \forall i \in [1, N], j \in [1, M]$.
8:     Return.
9: **end if**
10: Find the groups Λ_k^W, s.t. $\forall g_x \in \Lambda_k^W$, $\delta_{i_k, \sigma(i_k)} = \delta_{x, \sigma(x)}$.
11: Find the sellers Λ_k^S, s.t. $\forall s_y \in \Lambda_k^S$, $S_{j_k} = S_y$.
12: X_i is the sublist of the first i groups in X, and Y_j is the sublist of the first j sellers in Y.
13: Find X_k, Y_k, s.t. $|M(X_{k-1}, Y_{k-1})|$ is maximal, where $g_i \in \Lambda_k^W$ and $s_j \in \Lambda_k^S$ can be in any orders.
14: $G_W = \{g_x | \forall g_x \in M(X_{k-1}, Y_{k-1})\}$, $S_W = \{s_y | \forall s_y \in M(X_{k-1}, Y_{k-1})\}$.
15: $p^b = \delta_{i_k, \sigma(i_k)}$, $p^s = R_{j_k}$.
16: **for all** $i \in [1, L]$ **do**
17:     **if** Group $g_i \in G_W$ **then**
18:         For any buyer $k \in g_i$, $p_k^b = p^b / |g_i|$.
19:     **end if**
20: **end for**
21: **for all** $j \in [1, M]$ **do**
22:     **if** Seller $j \in S_W$ **then**
23:         $p_j^s = p^s$.
24:     **end if**
25: **end for**

Winner and Price Determination   In the winner and price determination process, the auctioneer finds the largest k such that the kth group's bid for the kth spectrum is no less than the kth seller's ask, $\delta_{i_k, j_k} \geq R_{j_k}$ (as each seller only owns one spectrum, we suppress the ask vector R_i as a single value). Instead of directly using the pair (g_{i_k}, s_{j_k}) to determine winners, we make one step further. Observed that there may be multiple equal g_{i_k}, s_{j_k} pairs, the auction outcome may be affected by the order of bids after sorting. So we check all the possible orders of bids equal to g_{i_k} or s_{j_k} and choose the one with the maximum matchings. The detailed algorithm is shown in Algorithm 6.

In line 10, the algorithm finds all group bids equal to $\delta_{i_k, \sigma(i_k)}$, and stores them in the set Λ_k^W. Actually, the order of group bids in Λ_k^W can be arbitrary. Similarly, Λ_k^S contains all sellers with asks equal to R_k. In line 13, the algorithm checks all the combinations of possible orders of groups in Λ_k^W and sellers in Λ_k^S to determine the number of matchings that can be achieved. The algorithm finally finds the maximum number of matchings for the given winner candidate sets G_C, S_C. $M(X, Y)$ denotes the set of matching induced by X and Y. From line 16 to line 25, the algorithm calculates the price for each buyer and the payment for each seller. The group price is equally shared by all the members in that group.

3.3.2 Illustrative Example

Figures 3.3– 3.5 shows a scenario with 6 buyers and 4 sellers. The bid vectors and buyers' interference relationships are shown in Fig. 3.3. For simplicity, we assume their interference relationships are the same across all spectrums. The sellers' asks are: $R_a = 7$, $R_b = 3$, $R_c = 4$ and $R_d = 2$ (Table 3.1).

Table 3.1 Members and bids for TAHES

Groups	a	b	c	d
$g_1 = \{A\}$	0	10	0	0
$g_2 = \{B, D\}$	4	3	0	5
$g_3 = \{C\}$	5	0	0	0
$g_4 = \{E, F\}$	0	0	5	0

After buyer grouping, the buyers form 4 groups with group bids shown in Table 3.2. The bids in italic font are chosen after the matching step. We can see that, there are three candidate bids with the same value 5. In the winner and price determination step, after sorting all the bids, according to Algorithm 6, in our scenario, $k = 3$, $\Lambda_k^W = \{g_2, g_3, g_4\}$, $\Lambda_k^S = \{c\}$. Then the algorithm checks all possible orders of groups in Λ_k^W and orders of sellers in Λ_k^S to determine the winners. In Fig. 3.4, there are two successful matchings; while in Fig. 3.5, there can be only one successful matching. We use the result in Fig. 3.4, which can be proved to have the maximum number of matchings. The price for the winning group is $\delta_{3,a} = 5$, the payment for the winning seller is $R_c = 4$. The price for winners in g_1 and g_2 will be shared by the members in these two groups. Therefore, $p_A = 5$, $p_B = p_D = 2.5$. The profit the auctioneer makes is $2 \times (5 - 4) = 2$.

3.3.3 Proofs of Economic Properties

In this section, we prove that TAHES is individually rational, budget-balanced and truthful.

Proposition 7 *TAHES is individually rational.*

Proof For each buyer i in winning group g_j,

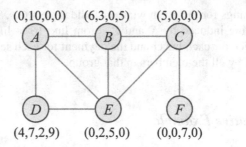

Fig. 3.3 An illustrative example: interference relationship of 6 buyers

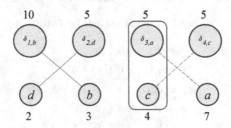

Fig. 3.4 An illustrative example: matching result 1

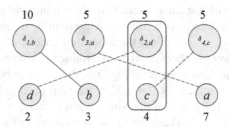

Fig. 3.5 An illustrative example: matching result 2

$$p_i^b = \frac{p^b}{|g_j|} \le \frac{\delta_{j,\sigma(j)}}{|g_j|} \le \frac{|g_j| \cdot b_{i,\sigma(j)}}{|g_j|} = b_{i,\sigma(j)}. \tag{3.6}$$

For each winning seller j, $p_j^s = p^s \ge R_j$.

Proposition 8 *TAHES is budget-balanced.*

Proof According to the sorting in Algorithm 6, we have $p^b \ge p^s$ and $|G_W| = |S_W|$, therefore the budget for the auctioneer is $|G_W| \times p^b - |S_W| \times p^s \ge 0$

Proposition 9   *TAHES is truthful.*

To prove Proposition 9, we first show that the auction result for buyer $i \in g_j$ is only related to the bid $b_{i,\sigma(j)}$. Then we show that the winner determination is monotonic and the pricing is bid-independent, such that for any buyer i or seller j, it cannot obtain higher utility by bidding untruthfully.

Lemma 1   *The auction result only depends on the buyers' bids for the assigned spectrum after spectrum matching.*

Proof   Both the grouping and the matching steps are bid-independent. After grouping, buyer i is in group g_j. After the matching step, only the group bid $g_{j,\sigma(j)}$ is considered in the winning and price determination procedure. $g_{j,\sigma(j)}$ is only related with buyer's bid $b_{i,\sigma(j)}$.

Now we only consider $b_{i,\sigma(j)}$ for buyer i.

Lemma 2   *The position of k remains the same for all combinations of orders of Λ_k^W and Λ_k^S.*

Proof   Since the group bids of groups in set Λ_k^W are equal and the seller bids of sellers in set Λ_k^S are also equal. The order of elements in Λ_k^W and Λ_k^S do not change k.

Lemma 3   *Given other buyers' bids $B\backslash\{-i\}$ for $s_{\sigma(j)}$, if buyer i wins in the auction, it also wins by bidding higher, that is, $b'_{i,\sigma(j)} > b_{i,\sigma(j)}$.*

Proof   If $b_{i,\sigma(j)}$ is not the minimum bid in group g_j, $b'_{i,\sigma(j)}$ will not change the group bid.

If $b_{i,\sigma(j)}$ is the minimum bid in group g_j, $b'_{i,\sigma(j)}$ will increase the group bid. Since the spectrum matching is independent of buyers' bids, the buyer group g_j will still be matched to the same spectrum. As the group bid increases, during the winning and price determination process, the buyer group g_j can still win the spectrum $s_{\sigma(j)}$.

Lemma 4   *Given other sellers' asks $R\backslash\{j\}$, if seller j wins in the auction, it also wins by asking $R'_j < R_j$.*

Proof   Since the spectrum matching is independent of sellers' asks, seller j should be matched to the same buyer group when asking lower. Since his ask decreases, during the winning and price determination process, seller j's spectrum can still be won by the buyer group $g_{\sigma^{-1}(j),j}$.

Lemma 5   *Given other buyers' bids $B\backslash\{-i\}$, if buyer $i \in g_j$ wins the auction by bidding $b'_{i,\sigma(j)}$ and $b_{i,\sigma(i)}$, the prices charged to i in both cases are the same.*

Proof   According to Lemma 2 and 3, increasing a winning buyer's bid will not change the auction results. It will not change k. Since the price is only dependent on k, the prices charged are the same.

Lemma 6   *Given other sellers' asks $R\backslash\{-j\}$, if seller j wins the auction by bidding R'_j and R_j, the prices paid to seller j are the same.*

Proof According to Lemma 2 and 4, decreasing a winning seller's bid will not change the auction results or k. Similarly with Lemma 5, the prices charged are the same.

Lemma 7 *TAHES is truthful for buyers.*

Proof We need to prove that a buyer $i \in g_j$ cannot increase his utility by bidding untruthfully. In other words, when buyer i bids $b'_{i,\sigma(j)} \neq v_{i,\sigma(j)}$, $U_i^{b'} \leq U_i^{b}$. Refer to the four possible auction results in Table 2.1.

- Case I. $U_i^{b'} = U_i^{b} = 0$.
- Case II. $U_i^{b'} = 0$, and $U_i^{b} \geq 0$ according to the property of individual rationality.
- Case III. According to Lemma 5, $U_i^{b'} = U_i^{b} > 0$.
- Case IV. According to Lemma 3, it happens only when $b'_{i,\sigma(j)} > b_{i,\sigma(j)} = v_{i,\sigma(j)}$. Since buyer i wins the auction by bidding higher, it must be true that buyer i's bid is the lowest bid in group g_j when bidding truthfully, and $\delta_{j,\sigma(j)} = |g_j| \cdot v_{i,\sigma(j)}$. Since buyer i wins when bidding untruthfully and loses when bidding truthfully, the group bid should satisfy $\delta'_{j,\sigma(j)} \geq \delta_{k,\sigma(k)} \geq \delta_{j,\sigma(j)}$, in which k is determined by Algorithm 6. The price paid by buyer i when bidding untruthfully is $p_i^{b} \geq \delta_{k,\sigma(k)}/|g_j| \geq \delta_{j,\sigma(j)}/|g_j| = v_{i,\sigma(j)}$. Therefore, $U_i^{b'} \leq U_i^{b} = 0$.

According to Lemma 1, a buyer cannot improve its utility by submitting any bid vector other than his true valuation vector.

Similarly, we can prove the following lemma on the truthfulness for sellers.

Lemma 8 *TAHES is truthful for sellers.*

Lemma 7 and Lemma 8 together prove the proposition 9 that TAHES is truthful.

3.4 Multiple Item Heterogeneous Spectrum Auction

In this section, we introduce a truthful multiple item heterogeneous spectrum auction mechanism TAMES [15], which allows each seller and buyer to trade multiple spectrums. We assume that buyers do not lie about their spectrum demand. This is because, if a buyer lies about his spectrum demand as $d'_i > d_i$, he may end up paying for more spectrums than he actually needs; if a buyer lies about his spectrum demand as $d'_i < d_i$, he never has the chance to win enough spectrums as he wants to. Therefore, the buyers are discouraged to be untruthful about their spectrum demands. In this case, we only focus on the truthfulness regarding the sellers' asks and the buyers' bids.

3.4.1 Auction Mechanism Design

TAMES consists of two steps: grouping and matching, and winner and price determination.

Grouping and Matching We use $G_j = (V_j, E_j)$ to denote the interference graph of $s_j, j \in [1, K]$. V_j is the set of buyers to whom spectrum s_j is available. G_j is constructed based on the transmission range of s_j. If the transmission range is long, more buyers will interfere with each other and there will be more edges in the interference graph. The group size shrinks because the size of independent set decreases. Consequently, the group bid decreases no matter how the group bid is decided under the constraint of individual rationality, making it harder for a group to win a spectrum. TAMES uses sequential grouping to mitigate the problem. The algorithm for sequential grouping is shown in Algorithm 7. The grouping process is performed by the auctioneer. It only depends on the buyers' interference relationships but not on their bids. After grouping, each group is matched to a specific spectrum.

- First, create dummies for the buyers with multiple demands. Every dummy inherits the interference relationships and the bidding profile of the original buyer. There is an edge between any two dummies of the same buyer so that they will not be grouped together.
- Then, sort the spectrums in non-descending frequency (non-ascending transmission range) order as $s_1, s_2, \ldots, s_K$, and sequentially group buyers for each spectrum. For spectrum s_j, we construct the interference graph G_j and find an independent set on the graph. This group is matched to spectrum s_j. In fact, it is not necessary to construct interference graph from scratch every time. We only have to (1) eliminate vertices of already grouped buyers and corresponding edges; (2) remove edges that no longer exist since the transmission range decreases for a spectrum with higher frequency. The auctioneer can choose different independent set selection policies for different purposes, e.g., maximal independent set may increase spectrum reusability, random picking independent set may ensure fairness.
- Finally, calculate the group bid as the lowest bid in the group for s_j multiplies the group size minus one.

Winner and Price Determination The procedure of winner and price determination is shown in Algorithm 5. After the grouping and matching, we compare each group bid with the seller's ask for the matched spectrum. If the group bid is higher than the ask, the auctioneer assigns the corresponding spectrum to every member in the group except the one with the lowest bid. The auctioneer then charges the group by the group bid, which will be split equally among all winners, i.e., each winner pays the same amount as the lowest bid of the group. The group member with the lowest bid is assigned nothing and pays nothing. If there are more than one group members who have the lowest bid, randomly pick one to be the "loser". The auctioneer will pay the seller of the spectrum by the group bid of the matched group.

Algorithm 7 Grouping and matching in TAMES

1: Unassigned buyer set V.
2: Create d_i dummies for buyer i if $d_i > 1$.
3: Delete the original buyer and insert the dummies in V.
4: Sort the spectrums in non-descending frequency order as $s_1, s_2, \cdots, s_K$.
5: **for all** $s_j, j = 1, 2, ..., K$ **do**
6:    **if** V is non-empty **then**
7:       V_j is the set of unassigned buyers for whom s_j is available.
8:       Construct interference graph based on s_j for buyers in V_j.
9:       Find one independent set g_j, and match g_j to s_j.
10:      The group bid of g_j is

$$\Phi_j = \min_{i \in g_j} b_{i,j} \times (|g_j| - 1), \tag{3.7}$$

in which $|g_j|$ is the number of buyers in the group;
11:      Eliminate members in g_j from V.
12:   **end if**
13: **end for**

Algorithm 8 Winner and price determination in TAMES

1: **for all** $g_j, j = 1, 2, ...$ **do**
2:    **if** $\Phi_j > r_{i,j}$ (i is the seller who owns s_j) **then**
3:       Group members in g_j except the the lowest-bid one are assigned spectrum s_j.
4:       Group g_j is charged by the group bid Φ_j.
5:       Buyer $l \in g_j, l \neq \arg_k \min_{k \in g_j} b_{k,j}$ is a winner of s_j, and the payment is:

$$p_{l,j}^b = \frac{\Phi_j}{|g_j| - 1} = \min_{k \in g_j} b_{k,j} \tag{3.8}$$

6:       Buyer $l = \arg_k \min_{k \in g_j} b_{k,j}$ loses, and pays nothing.
7:       Seller i is paid by the group bid.

$$p_{i,j}^s = \Phi_j. \tag{3.9}$$

8:    **else**
9:       All buyers in g_j lose and pay nothing.
10:      The seller i loses and is paid nothing.
11:   **end if**
12: **end for**

3.4.2 Illustrative Example

Figures 3.6, 3.7 and 3.8 show how TAMES works. There are 6 buyers. Buyer D's demand is 2, so we create two dummies D_1 and D_2. There are 2 sellers. One seller has 2 spectrums a and b with asks $r_{1,a} = 6, r_{1,b} = 3$. The other seller has one spectrum c with ask $r_{2,c} = 2$. We assume that the frequencies of the spectrums are sorted in the increasing order as a, b, c. We also assume that each spectrum is available to all buyers. The bidding profile of each buyer is shown in the figure.

In the first iteration, interference graph on spectrum a is shown in Fig. 3.6. Buyer A, E form g_1 with group bid 5, which equals the bid of buyer E for a. In the second iteration as shown in Fig. 3.7, A, E are eliminated from the graph, along with the edges between (B, D_2) and (C, D_1). This is because these two pairs no longer interfere with each other on spectrum b, which has a higher frequency and shorter transmission range than spectrum a. Buyer B, D_2 form group g_2 with group bid 4. In the third iteration as shown in Fig. 3.8, B, D_2 and edge (D_1, F) are eliminated, leaving C, D_1, F forming g_3 with group bid 4.

Table 3.2 Members, bids and asks for TAMES

Groups	a	b	c
$g_1 = \{A, E\}$	5	None	None
$g_2 = \{B, D_2\}$	None	4	None
$g_3 = \{C, D_1, F\}$	None	None	4
Ask	6	3	2

After comparing group bids with the asks of the sellers, the auctioneer assigns spectrum b to D_2. D pays 4 and the seller of b gets payment of 4. C, D_1 are assigned spectrum c and pay 2 respectively. The seller who owns c gets payment of 4. D wins two spectrums via his two dummies, and pays a total of 8. Other buyers lose in the auction.

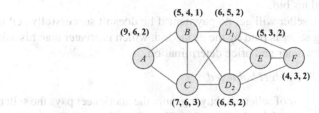

Fig. 3.6 Interference graph on a

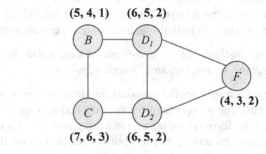

Fig. 3.7 Interference graph on b

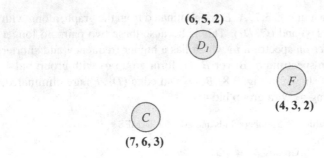

Fig. 3.8 Interference graph on c

3.4.3 Proofs of Economic Properties

In this section, we give proofs for the individual rationality, budget balance, and truthfulness of TAMES.

Proposition 10 *TAMES is individually rational.*

Proof On the buyers' side, a buyer will be charged nothing if he loses on a spectrum. Each winning buyer is charged by the price of the lowest bid in his group, so his payment will not exceed his bid.

On the sellers' side, a seller will get no payment if he doesn't successfully sell a spectrum. Each winning seller is paid by the group bid, which is greater than his ask according to the rule of winner and price determination.

Proposition 11 *TAMES is budget balanced.*

Proof For each winning pair of seller and buyer group, the auctioneer pays the seller and charges the buyer group both by the group bid. Therefore, the auctioneer's utility is zero, which is non-negative.

On the buyers' side, we prove the truthfulness from two aspects. We first prove that a buyer's bid for a specific spectrum does not affect the winning result of another spectrum (inter-spectrum truthfulness). In other words, for buyer i, $b_{i,m}$ and $w_{i,n}^b$, $m \neq n$ are irrelevant. Then, we prove that a buyer cannot misreport his bid for a specific spectrum to achieve higher utility from that spectrum (intra-spectrum truthfulness).

Lemma 9 *Inter-spectrum truthfulness. A buyer cannot manipulate his bid for a specific spectrum to affect the winning result of another spectrum.*

Proof The buyers have no control over the grouping and matching process. During sequential grouping process, the group members are matched to a spectrum by the auctioneer. The auctioneer performs grouping and matching process only depending on the interference relationships among buyers, but not depending on the buyers' bids.

The group bid only relies on the group members' bids for the matched spectrum, independent of their bids for other spectrums. That is to say, for buyer i, $b_{i,m}$ and $w_{i,n}^b$, $m \neq n$ are independent.

Lemma 10 *Intra-spectrum truthfulness. A buyer cannot misreport his bid for a specific spectrum to increase his utility gained from that spectrum.*

Proof Assume that buyer i's group is matched to spectrum j. Buyer i's utility gained from s_i is as follows.

$$U_{i,j}^b = w_{i,j}^b \times v_{i,j}^b - p_{i,j}^b$$

Let $w_{i,j}^b$ and $U_{i,j}^b$ denote the winning result and utility when buyer i bids truthfully. Let $w_{i,j}^{b'}$ and $U_{i,j}^{b'}$ denote the winning result and utility when buyer i bids untruthfully. When buyer i bids truthfully, his bid equals $v_{i,j}^b$. We will prove that $U_{i,j}^b \geq U_{i,j}^{b'}$ is always true.

We discuss the four cases listed in Table 2.1.

- CASE I. $w_{i,j}^b = 0, w_{i,j}^{b'} = 0$. In this case, $U_{i,j}^b = U_{i,j}^{b'} = 0$.
- CASE II. $w_{i,j}^b = 1, w_{i,j}^{b'} = 0$. In this case, $U_{i,j}^{b'} = 0$, and $U_{i,j}^b \geq 0$ due to individual rationality.
- CASE III. $w_{i,j}^b = 1, w_{i,j}^{b'} = 1$. Since buyer i wins, his bid is not the minimum or if his bid is the minimum, there will be other group member(s) whose bid(s) is(are) also minimum. Therefore, buyer i's payment equals the minimum bid in the group, which will not be affected by buyer i's bid. So we have $U_{i,j}^b = U_{i,j}^{b'}$.
- CASE IV. $w_{i,j}^b = 0, w_{i,j}^{b'} = 1$. This can only happen when buyer's truthful bid is the minimum in the group, and by increasing his bid, his bid is no longer the minimum. When buyer i bids untruthfully and wins, his payment $p_{i,j}^{b'} > v_{i,j}^b$. Therefore, $U_{i,j}^{b'} = v_{i,j}^b - p_{i,j}^{b'} < 0 = U_{i,j}^b$.

In summary, buyer i does not have incentive to be untruthful since he cannot gain higher utility by misreporting his true valuation.

Now we prove the truthfulness on the sellers' side.

Lemma 11 *TAMES is truthful for sellers.*

Proof It can be easily proved that a seller cannot misreport his true valuation for a specific spectrum to affect the winning result of another spectrum. This is because his ask for one spectrum does not affect the winning result of another spectrum. Now we prove that the seller cannot misreport the ask for a specific spectrum to increase his utility obtained from that spectrum. We assume that seller i owns spectrum j and his ask for s_j is $r_{i,j}$. Seller i's utility gained from s_j is as follows.

$$U_{i,j}^s = p_{i,j}^s - w_{i,j}^s \times v_{i,j}^s$$

Let $w_{i,j}^s$ and $U_{i,j}^s$ denote the winning result and utility when seller i asks truthfully. Let $w_{i,j}^{s'}$ and $U_{i,j}^{s'}$ denote the winning result and utility when seller i asks untruthfully.

When seller i asks truthfully, his ask equals $v_{i,j}^s$. We assume that when seller i asks untruthfully, his ask is $r_{i,j}'$. We will prove that $U_{i,j}^s \geq U_{i,j}^{s'}$ is always true.

We discuss the four cases listed in Table 2.1.

- CASE I. $w_{i,j}^s = 0, w_{i,j}^{s'} = 0$. In this case, $U_{i,j}^s = U_{i,j}^{s'} = 0$.
- CASE II. $w_{i,j}^s = 1, w_{i,j}^{s'} = 0$. In this case, $U_{i,j}^{s'} = 0$, $U_{i,j}^s \geq 0$ due to individual rationality.
- CASE III. $w_{i,j}^s = 1, w_{i,j}^{s'} = 1$. In this case, the seller is paid the group bid, independent of his ask. $U_{i,j}^s = U_{i,j}^{s'}$.
- CASE IV. $w_{i,j}^s = 0, w_{i,j}^{s'} = 1$. It can only happen when $r_{i,j}' < \Phi_j < r_{i,j} = v_{i,j}^s$, in which Φ_j is the group bid. In this case, $U_{i,j}^{s'} = \Phi_j - v_{i,j}^s < 0 = U_{i,j}^s$.

In summary, the seller cannot improve its utility gained from a certain spectrum by misreporting the ask for that spectrum.

Proposition 12 *TAMES is truthful.*

Proof By combining lemmas 9, 10 and 11, we can establish that TAMES is truthful on both buyers' and sellers' side.

3.4.4 Spectrum Continuity

In this section, we tweak the grouping algorithm to improve the spectrum continuity, and verify by simulation that the modified grouping algorithm increases the chance that buyers who win multiple spectrums get continuous spectrums.

Buyers with multiple spectrum demands prefer to have continuous spectrums mainly because of two reasons. First, some devices can only work with continuous spectrums. Second, even if multiple devices are involved, it is still more desirable to have continuous spectrums due to easy management. However, during the grouping process, dummies of the same buyer are assigned randomly to different groups for different spectrums. So the auction results do not guarantee the spectrum continuity of the buyers who win multiple spectrums.

A naive solution is to let buyers exchange spectrums after the auction. This is feasible if buyers have a homogenous valuation for all spectrums. However, since buyers have heterogeneous valuations, they are unwilling to exchange high-valued spectrums for low-valued spectrums. Also, the prices for different spectrums are different. It is impossible to carry out such spectrum exchange to accommodate buyers' need for continuous spectrums.

In order to address the problem of spectrum discontinuity, we make use of a specific independent set algorithm in [43]. The basic idea is: when seeking for group g_j for spectrum s_j, if a node v has a brother dummy (we refer to dummies of the same buyer as the brother dummies) in g_{j-1}, we increase the probability that v is put in g_j. But such "bonus" will be deprived when v is considered for the next group g_{j+1} for fairness.

To quantify the degree of spectrum continuity, we define a Continuity Index CI. CI is defined as the sum of intervals between two consecutive spectrums won by a buyer divided by the number of spectrums obtained. Let $s_{i_1}, s_{i_2}, ..., s_{i_m}$ denote the spectrums won by buyer i.

$$CI_i = \frac{\sum_{k=2}^{m} (i_k - i_{k-1} - 1)}{m - 1} \tag{3.10}$$

The higher the continuity index CI is, the less continuous all the spectrums are. Table 3.3 shows that CT algorithm significantly reduces the continuity index, with little sacrifice of sellers' revenue and spectrum reusability.

Table 3.3 Comparison of CT and non-CT algorithms

Number of spectrums		10	15	20
Continuity Index	Non-CT	35.94	78.19	104.13
	CT	16.61	34.17	46.59
Revenue	Non-CT	22.27	34.59	42.53
	CT	20.56	30.49	40.21
Continuity	Non-CT	12.26	9.76	7.98
	CT	11.03	9.39	7.20

Chapter 4
Dynamic Spectrum Auction

Static spectrum auction, when applied to dynamic spectrum allocation, will cause potential utility loss. For example, in Fig. 4.1, suppose that buyer A arrives at the first time stage, requesting for 3 time slots, and his true valuation is 1 per time slot. Later at the second time stage, buyer B and buyer C arrive, each requesting for 2 time slots, and their true valuations are both 1 per time slots. If the auctioneer allocates the spectrum to buyer A at the first time stage, buyer A gets utility of 3. However, if the auctioneer rejects buyer A in the first time stage and allocates the spectrum to buyer B and C simultaneously at the second time stage, they get a total utility of 4. Although the spectrum is idle in the first time stage, overall, the spectrum utilization of the second option is higher.

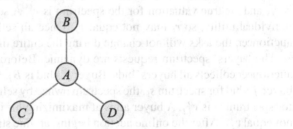

Fig. 4.1: A motivated example

Most of the existing spectrum auction mechanisms are static, which will incur utility loss in case of dynamic spectrum allocation. Online spectrum auction is considered in [63] but it is assumed that the interference graph is complete. While such an assumption makes the auction design easier, it does not capture the most important feature of spectrums, that is, spectrum reusability. In consequence, the spectrum utilization is low.

To design an online auction mechanism, we need to take into consideration the sporadic nature of spectrum requests and buyers' geographic feature. The challenges are two-folds:

© The Author(s) 2015 37
Y. Chen, Q. Zhang, *Dynamic Spectrum Auction in Wireless Communication*,
SpringerBriefs in Electrical and Computer Engineering, DOI 10.1007/978-3-319-14030-8_4

- *How to characterize the interference harmfulness of a certain buyer?*
  This is difficult because, although a buyer may interfere with many other buyers, it is possible that one of his interfering neighbors interferes with even more buyers.
- *How to decide whether to allocate the spectrum to a certain buyer based on the knowledge of the spectrum requests in the current time stage?*
  This is difficult because spectrum requests are random and it is hard to predict whether a buyer's interfering neighbors will request spectrum in later time stages.

Apart from the above challenges, the online auction mechanism should also be economic-robust, that is, truthful, individual rational and budget balanced for the auctioneer.

In this chapter, we introduce LOTUS [14], a location-aware online double spectrum auction, which aims at exploiting location heterogeneity to improve spectrum utilization while guaranteeing economic robustness.

4.1 Modeling Online Spectrum Auction

We assume there are M sellers and N buyers. Each seller has one spectrum. We assume that all spectrums are available during the entire time stages from 1 to T. If some spectrums are unavailable in certain time stages, the problem can be easily fixed by removing the spectrums from all buyers' available spectrum sets. Before the start of the auction, the auctioneer collects the asks of all sellers. Seller i's ask is r_i, and his true valuation for the spectrum is v_i^s. A seller aims at maximizing his individual utility, so r_i may not equal v_i^s. Once all sellers submit their asks to the auctioneer, the asks will not change during the entire time stages $[1, T]$.

The buyers' spectrum requests are dynamic. Before the start of the auction, the auctioneer collects all buyers' bids. Buyer j's bid is $B_j = (b_{j,1}, b_{j,2}, \ldots, b_{j,M})$. $b_{j,i}$ is buyer j's bid for spectrum s_i, the spectrum owned by seller i. Buyer j's true valuation for spectrum s_i is $v_{j,i}^b$. A buyer aims at maximizing his individual utility, so $b_{j,i}$ may not equal $v_{j,i}$. After the online auction begins, at time stage t, buyer j will submit an instant spectrum request, specifying the number of requested spectrums d_j and the number of desired time slots t_j to the auctioneer. The number of desired time slots t_j means that the buyer wants to occupy the spectrum during $t \sim t + t_j$. We assume that a buyer's time slot request is strict. In other words, buyer i either accepts t_i time slots or no time slots. We assume that a buyer's spectrum request is not strict. That is to say, buyer i accepts a number of spectrums in $[0, d_j]$. If a buyer does not want any spectrum at time stage t, his demand is zero. To make the following analysis clearer, we give formal definitions of interfering neighbors and interference degree as follows.

Definition 2 *Interfering neighbors.* The interfering neighbors of buyer j is the set of buyers who interfere with buyer j, i.e., the nodes who share an edge in the interference graph with buyer j.

Definition 3 *Interference degree.* The interference degree of buyer j is the number of his interfering neighbors, that is, the degree of buyer j's node in the interference graph.

We use $N(j)$ to denote the set of interfering neighbors of buyer j and θ_j to denote his interference degree.

The auctioneer knows the spatial availability of each spectrum to decide available spectrum set for each buyer. The buyers give the auctioneer their location information for the auctioneer to compute the interference discount, which we will introduce later. We assume that the objective of the auctioneer is to maintain a balanced budget at each time slot, that is, $U_t^a \geq 0, t = 1, 2, \cdots, T$.

At each time stage t, all buyers submit their spectrum requests to the auctioneer, then the auctioneer decides the spectrum allocation and the corresponding payments. The auction results are represented by a 3-dimensional matrix $X_{i,j,t}, i \in [1, M], j \in [1, N], t \in [1, T]$. At time stage t, if the auctioneer decides to allocate the spectrum s_i to buyer j for t_j time slots, then $x_{i,j,t} = 1, x_{i,j,t+1} = 1, \ldots, x_{i,j,t+t_j} = 1$; otherwise, $x_{i,j,t} = 0$.

At time stage t, if seller i's spectrum is assigned to at least one buyer, the auctioneer will pay seller i $p_{i,t}^s$; otherwise, the auctioneer will pay seller i nothing. The utility of seller i is his payment from the auctioneer minus his true valuation.

$$
U_{i,t}^s = \begin{cases} p_{i,t}^s - v_i^s & \text{if } \sum_j x_{i,j,t} > 0 \\ 0 & \text{otherwise} \end{cases} \tag{4.1}
$$

At time stage t, if buyer i is assigned spectrum j, he will be charged $p_{i,j,t}^b$. The utility of buyer j is his true valuation minus his payment to the auctioneer.

$$
U_{i,t}^b = \sum_j \left[(v_{i,j}^b - p_{i,j,t}^b) \times x_{j,i,t} \right]. \tag{4.2}
$$

The auctioneer's budget at time t is the payment obtained from all the buyers minus the payment to all the sellers.

$$
U_t^a = \sum_{j,i} p_{j,i,t}^b - \sum_i p_{i,t}^s \tag{4.3}
$$

To estimate the influence of the spectrum allocation in the current time stage on the spectrum allocation in the following time stage, we make the following assumptions.

1. Buyers' spectrum requests are Independent and Identically Distributed (IID) random variables. More specifically, buyer's spectrum requests follow Poisson distribution with arrival rate λ.
2. Buyers' bids are IID random variables, and the distribution is known by the auctioneer.
3. The number of requested time slots of each buyer is IID random variables, and the distribution is known by the auctioneer.

Assumption 1 is a common assumption for traffic in wireless network ([37, 63]). Assumption 2 is reasonable as the auctioneer can estimate the trend of buyers' bids either from the bidding history or by evaluating the value of the spectrum in an open market. Assumption 3 is reasonable as the time for performing common tasks in wireless communication can be estimated based on historical data.

We define truthfulness, individual rationality and budget balance for online spectrum auction as follows.

- Truthfulness. For the online spectrum auction, we have to consider the truthfulness at each time stage. Let $\widetilde{U}_{i,t}^s$ and $U_{i,t}^s$ denote the utility of seller i when he asks truthfully and untruthfully respectively. For an online auction to be truthful, $U_{i,t}^s \geq \widetilde{U}_{i,t}^s$ always holds. As for the buyers, the truthfulness includes three aspects: (1) buyers bid their true valuations; (2) buyers submit their true numbers of requested spectrums; (3) buyers submit their true numbers of desired time slots. We assume that the buyers do not lie about their numbers of requested spectrums and desired time slots. The reason is: if a buyer submits a fewer number of spectrums or time slots, he cannot finish his task, and will get negative utility; if a buyer submits a larger number of spectrums or time slots, he has to make extra payments, and will get negative utility. Therefore, we only consider the truthfulness regarding buyers' bidding price. Assume that buyer j gets $\widetilde{U}_{j,t}^b$ and $U_{j,t}^b$ when he bids truthfully and untruthfully respectively. For an online auction to be truthful, $U_{j,t}^b \geq \widetilde{U}_{j,t}^b$ always holds.
- Individual rationality. For the online spectrum auction to be individual rational, at each time stage, the seller is paid more than his ask, i.e., $p_{i,t}^s \geq r_i$, and the buyer pays less than his bid, i.e., $p_{j,i,t}^b \leq b_{j,i}$.
- Budget balance. For the online spectrum auction to be individual rational, at each time stage, the auctioneer gets non-negative payment, i.e., $U_t^a \geq 0$.

The objective of the online spectrum auction mechanism can be described as: Given the temporal and spatial features of the buyers' spectrum requests, how to find a truthful, individually rational and budget balanced double auction mechanism to dynamically allocate the spectrum at each time stage?

4.2 Interference Discount

In this section, we first introduce two factors that can describe the interference harmfulness of a certain buyer. Then we combine the two factors to compute the interference discount.

As shown in Fig. 4.1, if we simply run an existing static auction mechanism at each time stage where the auctioneer considers only the spectrum request at the current time stage, it will incur utility loss. The reason is that, if buyer j wins a spectrum i at time t, he will occupy the spectrum from $t \sim t + t_j$. His interfering neighbors cannot use spectrum i if they request at a later time stage during $t + 1 \sim t + t_j$, even if they bid higher prices, or can reuse the spectrum more efficiently. An important observation is that, granting the spectrum request of a buyer who is located in a *critical* place and interferes with a lot of other buyers, may generate high potential utility loss. Therefore, we have to take into consideration buyers' spatial features, especially their interference relationships, in order to improve spectrum utilization.

We analyze the interference harmfulness of a buyer from two aspects. First, we compare the interference degree of a buyer with the average interference degree of his interfering neighbors. Second, we consider the reusability efficiency of a buyer's interfering neighbors. We combine the two aspects together to decide the opportunity cost of allocating the spectrum to a certain buyer.

4.2.1 Comparison of Interference Degree

For buyer j, there exists two spectrum allocation options: to allocate the spectrum to buyer j; or to allocate the spectrum to one of his interfering neighbors. Another possible choice is not to assign the spectrum to either buyer i or any of his interfering neighbors, but it is unreasonable since we can always improve spectrum utilization by allocating the spectrum to either party without violating the interference constraint. In the former case, none of the buyers in $N(j)$ can use the same spectrum. In the later case, none of the buyers who are the interfering neighbors of the buyers in $N(j)$ can use the same spectrum. Such usage exclusivity has a ripple effect, which may extend to the entire interference graph. To make the problem tractable, we only consider 3 layers, from buyer j to his interfering neighbors in $N(j)$, and then to the interfering neighbors of those in $N(j)$. Let δ_j denote the ratio of the interference degree of buyer j to the average interference degree of his interfering neighbors.

$$\delta_j = \frac{\theta_j}{(\sum_{k \in N(j)} \theta_k)/\theta_j} = \frac{\theta_j^2}{\sum_{k \in N(j)} \theta_k} \qquad (4.4)$$

If δ_j is large, to allocate the spectrum to buyer j will have negative effect on more buyers than to allocate the spectrum to the buyers in $N(j)$, vice versa. Hence, δ_j shows the harmfulness of a buyer in terms of his interference degree.

The δ_j in different cases is shown in Fig. 4.2. The central node denotes buyer j. In Fig. 4.2a, the buyer's interfering neighbors have smaller average interference degree, so $\delta_j > 1$, which means that buyer j is quite interference harmful. In Fig. 4.2b, the buyer's interfering neighbors have exactly the same average interference degree, so $\delta_j = 1$. In Fig. 4.2c, the buyer's interfering neighbors have larger average interference degree, so $\delta_j < 1$, which means that buyer j may be a good candidate for spectrum allocation than his interfering neighbors.

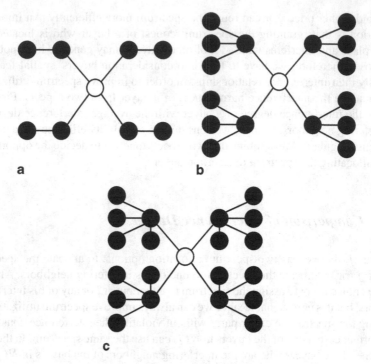

Fig. 4.2: Comparison of interference degree

4.2.2 *Reusability Efficiency of Interfering Neighbors*

It is not enough to only consider the interference degree. For example, as shown in Fig. 4.3, the central node denotes buyer j. By definition, $\delta_j = 4/3$ for all three cases. However, in Fig. 4.3a, since the interfering neighbors can reuse the same spectrum, to allocate the spectrum to buyer j causes more utility loss than the case in Fig. 4.3c, where none of the interfering neighbors can reuse the same spectrum. Therefore, we have to consider the spectrum reusability efficiency of a buyer's interfering neighbors.

To determine the reusability efficiency among a set of buyers is not easy. For instance, in Fig. 4.3b, some of the interfering neighbors of buyer j can reuse the same spectrum, but some cannot. One way is to compute the average reusability efficiency by considering all possible subsets of buyer j's interfering neighbors. However, when the number of buyers increases, the number of subsets increases exponentially. For simplification, we only consider the maximum reusability efficiency (denoted by σ_i), that is, the maximum number of buyers in $N(j)$ that can reuse the spectrum.

Figure 4.3a shows the case in which all the interfering neighbors are interference free, so $\sigma_j = 4$. In Fig. 4.3b, the maximum number of interfering neighbors who can reuse the spectrum is 2, so $\sigma_j = 2$. In Fig. 4.3c, which is a complete graph, $\sigma_j = 1$.

As we discussed before, to allocate the spectrum to buyer j causes the most utility loss in Fig. 4.3a as his interfering neighbors can reuse the spectrum most efficiently.

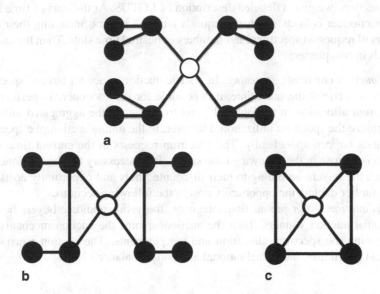

Fig. 4.3: Reuse degree of interference neighbors

4.2.3 Interference Discount

We define the *Interference Discount* as follows.

Definition 4 *Interference Discount.* Interference discount is the discount factor that marks down a buyer's bid by the comparison of interference degree and the reusability efficiency of his interfering neighbors. The interference discount of buyer j is as follows.

$$I_i = \frac{1}{\delta_i \sigma_i} \tag{4.5}$$

In other words, if buyer j bids $b_{j,i}$ for s_i, the auctioneer discounts the bid as $I_j \cdot b_{j,i}$ when screening the buyers for valid candidates of the auction. We assume that the interference relationships among the buyers are stable. So the interference discount of each buyer can be calculated offline, and maintained as a lookup table by the auctioneer.

4.3   Auction Mechanism Design

In this section, we give a detailed description of LOTUS. At the start of time stage t, the auctioneer collects spectrum requests from all buyers, including their bids, numbers of requested spectrums and numbers of desired time slots. Then the auction proceeds in two phases:

- *Pre-auction candidate screening.* In a static auction, since all buyers' spectrum requests arrive at the same time, it is possible for the auctioneer to perform the spectrum allocation in an "optimal" way to maximize the aggregated utility or maximize the spectrum utilization. However, in the online auction, the spectrum requests happen sporadically. The spectrum requests in the current time stage influence those in the following time stages. It is necessary for the auctioneer to screen the buyers according to their discounted bids and opportunity costs. We will further explain the opportunity cost in the following sections.
- *Main auction.* After pre-auction screening, the valid candidate buyers become potential auction winners. Then the auctioneer runs the auction mechanism to determine the spectrum allocation and the payments. The auction mechanism should be truthful, individual rational and budget balanced.

4.3.1   Pre-auction Candidate Screening

Now we explain how to calculate the opportunity cost of allocating the spectrum s_i to buyer j at time stage t. Before the auction starts, the auctioneer will compute the expected valuation matrix $Y_{N \times M \times T}$. Entry $y_{j,i,t}, j \in [1, N], i \in [1, M], t \in [1, T]$ denotes the expected value of allocating s_i to the interfering neighbors of buyer j from time stage t on. Let $\mathbf{Pr}(\tau)$ denote the probability that the number of desired time slots of a typical buyer is τ, $\bar{b}_{j,i} = \sum_{k \in N(j)} b_{k,i} I_k / |N(j)|$ denote the average discounted bids of the interfering neighbors in $N(j)$ for spectrum s_i. Then $y_{j,i,t}$ is calculated as follows:

$$y_{j,i,t} = \sum_{\tau=1}^{T-t} [\mathbf{Pr}(\tau)(\bar{b}_{j,i}\tau + y_{j,i,t+\tau})] \tag{4.6}$$

Y is an $N \times M \times T$ matrix and can be calculated using dynamic programming.

The opportunity cost of allocating s_i to the buyer j at time stage t for t_j time slots is

$$C_{j,i}(t, t_j) = y_{j,i,t} - y_{j,i,t+t_j}. \tag{4.7}$$

At time stage t, the auctioneer compares the opportunity cost $C_{j,i}(t, t_j)$ with the discounted bid $b_{j,i} * I_j$ to decide whether the buyer j is a valid candidate for s_i.

- If $C_{j,i}(t, t_j) > b_{j,i} * I_j * t_j$, meaning that buyer j does not bid high enough to compensate the potential spectrum utilization loss. Therefore, buyer j will not be considered as a valid candidate for winning s_i.

- If $C_{j,i}(t,t_j) \leq b_{j,i} * I_j * t_j$, meaning the buyer j's bid exceeds the potential spectrum utilization loss, the auctioneer considers him as a valid candidate for winning s_i.

Based on the pre-auction candidate screening results, the auctioneer can construct the available spectrum set for each buyer who submits the spectrum request at time stage t, as shown in Algorithm 9. Let $S_{j,t}$ denote the available spectrum set for buyer j at time stage t. Firstly, all the spectrums are put into $S_{j,t}$. Then, the spectrums that are occupied by the interfering neighbors of buyer j are removed from $S_{j,t}$. Finally, the spectrums for which buyer j is not a valid candidate are removed from $S_{j,t}$.

Algorithm 9 Available spectrum set construction in LOTUS

1: **for all** Time stage $t = 1, 2, ..., T$ **do**
2:     **for all** Buyer $j, j = 1, 2, ..., N$ **do**
3:         Put all the spectrums in $S_{j,t}$.
4:         **for all** $s_i, i = 1, 2, ..., M$ **do**
5:             **if** s_i is being occupied by any buyer in $N(j)$ **then**
6:                 Remove s_i from $S_{j,t}$.
7:             **end if**
8:             **if** $C_{j,i}(t,t_j) > b_{j,i} * I_j * t_j$ then
9:                 Remove s_i from $S_{j,t}$.
10:            **end if**
11:         **end for**
12:     **end for**
13: **end for**

After the pre-auction candidate screening, each buyer gets an available spectrum set. Then the auctioneer moves on to the main auction phase.

Algorithm 10 Grouping in LOTUS

1: **for all** Time stage $t = 1, 2, ..., T$ **do**
2:     $G_t = \{V_t, E_t\}$ is interference graph consisting of buyers who submit spectrum requests at time stage t.
3:     Add edges between any two buyers who do not have a common spectrum in their available spectrum sets.
4:     **for all** Buyer $j, j \in V_t$ **do**
5:         Create d_j dummies for buyer j if $d_j > 1$.
6:         Delete the original buyer j in G_t and insert the dummies in G_t.
7:     **end for**
8:     **for all** $s_i, i = 1, 2, ..., M$ **do**
9:         **if** V_t is non-empty **then**
10:             $V_{t,i}$ is the set of buyers for which s_i is available.
11:             Find one independent set g_i on $V_{t,i}$.
12:             Match g_i to s_i.
13:             Let k denote the buyer with minimum $(b_{k,i}t_k)$ in g_i. The group bid of g_i is

$$\Phi_i = \frac{b_{k,i}t_k}{\max\limits_{j \in g_i \setminus \{k\}} t_j} \times (|g_i| - 1)$$

        in which $|g_i|$ is the size of g_i.
14:             Remove g_i from V_t.
15:         **end if**
16:     **end for**
17: **end for**

4.3.2 Main Auction Algorithm

The spectrum allocation includes two parts: grouping, and winner and price determination.

Grouping Buyers who can reuse the same spectrums can be grouped together, and allocated the same spectrum in order to improve spectrum utilization. In the previous double auction mechanisms [26, 64, 72], it is assumed that all the spectrums are available to each buyer, so the grouping process can be easily done by partitioning the interference graph into multiple independent sets. However, this no longer works when different buyers have different available spectrum sets. Two buyers who do not have any common available spectrums cannot be grouped together even if they do not interfere with each other. To solve these problems, we use Algorithm 10 for grouping.

- Add virtual edges in the interference graph to make sure that buyers with no common available spectrum will not be grouped together.
- Create dummies to replace the original buyer. If the buyer requests d_j spectrums, then create d_j dummies. In this way, the buyer has d_j chances to be included in an independent set, and to win up to d_j spectrums.
- When computing the group bid, the buyer who bids the minimum bid is sacrificed in order to guarantee truthfulness.

Winner and Price Determination Winner and price determination is in Algorithm 11.

Algorithm 11 Winner and price determination in LOTUS

1: **for all** $g_i, i = 1, 2, ..., M$ **do**
2:     **if** $\Phi_i > r_i$ **then**
3:         All buyers in g_i except k win s_i.
4:         Group g_i pays a total amount of $\Phi_i \times \max_{j \in g_i \setminus \{k\}} t_j$, which is equally shared by the winning buyers (except k).
5:         Seller of s_i is paid Φ_i per time stage till $t + \max_{j \in g_i \setminus \{k\}} t_j$;
6:     **else**
7:         All buyers in group g_i lose and pay nothing.
8:         Seller of s_i is paid nothing.
9:     **end if**
10: **end for**

4.4 Proofs of Economic Properties

In this section, we prove that LOTUS is economic robust in terms of individual rationality, budget balance and truthfulness.

Individual Rationality Theorem 1 *LOTUS is individual rational.*

Proof Let's consider the winning seller i and the winning buyer group g_i.

- For seller i, the per time slot payment Φ_i is greater than r_i, and the spectrum occupancy is no more than $\max_{j \in g_i \setminus \{k\}} t_j$. Therefore, the seller i gets positive utility as calculated by (4.1).
- For a winning buyer j in g_i, the total payment is

$$\frac{\Phi_i \times \max_{j \in g_i \setminus \{k\}} t_j}{(|g_i| - 1)} = b_{k,i} t_k \le b_{j,i} t_j$$

Therefore, the total payment is less than the total bid of buyer j, and he can get positive utility as calculated by (4.2).

Budget Balance Theorem 2 *LOTUS is budget balanced.*

Proof Let's consider the payment to the winning seller i and the payment from the winning buyer group g_i. The auctioneer's budget is

$$\Phi_i \times \max_{j \in g_i \setminus \{k\}} t_j - \Phi_i \times \max_{j \in g_i \setminus \{k\}} t_j = 0$$

Therefore, the auctioneer's budget is always zero, which is a non-negative budget.

Truthfulness We only consider the truthfulness after the screening phase. We first prove the truthfulness on the buyers' side, then prove the truthfulness on the sellers' side. Since the buyers can bid for multiple spectrums, and can bid different prices for different spectrums, we also have to consider intra-spectrum truthfulness and inter-spectrum truthfulness.

Lemma 12 *Inter-spectrum truthfulness. A buyer cannot manipulate the bidding price for one spectrum to affect the winning result of another spectrum.*

Proof Similar to the proof of Lemma 9.

We give a slightly different proof for intra-spectrum truthfulness by proving the following two lemmas.

Lemma 13 *If buyer j wins spectrum s_i, he pays the same price regardless of his bid $b_{j,i}$ for s_i.*

Proof If buyer j wins in group g_i, he is not the one with the minimum bid and he will pay $b_{k,i} t_k t_j / \max_{j \in g_i \setminus \{k\}} t_j$, which is unaffected by $b_{j,i}$.

Lemma 14 *If buyer j wins spectrum s_i by bidding $b_{j,i}$, he will also win by bidding $b'_{j,i} > b_{j,i}$.*

Proof Since buyer j is a winning buyer in group g_i, $b_{j,i} t_j$ is not the minimum in group g_i. Therefore, the group bid will not be affected if $b_{j,i}$ increases to $b'_{j,i}$. Group g_i will still win spectrum s_i, and buyer j will still be a winning buyer in group g_i.

Now, we can leverage Lemma 13 and 14 to prove intra-spectrum truthfulness.

Lemma 15 *Intra-spectrum truthfulness. A buyer cannot manipulate the bidding price for one spectrum to affect the winning result of the spectrum.*

Proof Let's consider buyer j and spectrum s_i. Buyer j bids truthfully as $v_{j,i}^b$, and untruthfully as $b'_{j,i}$. We consider four possible auction results as shown in Table 2.1.

- Case I. Buyer j loses the spectrum s_i when he bids truthfully and untruthfully, and his utilities are both zero. Therefore, the buyer does not gain higher utility by being untruthful.
- Case II. Buyer j wins the spectrum s_i when he bids truthfully and loses the spectrum s_i when he bids untruthfully. In the former case, he achieves non-negative utility due to individual rationality. In the later case, he achieves zero utility. Therefore, the buyer has higher utility when being truthful.
- Case III. Buyer j wins the spectrum s_i when he bids truthfully and untruthfully, and he pays the same price according to Lemma 13. Since the buyer also gains the same valuation for using spectrum s_i, his utilities are the same when he bids truthfully and untruthfully.
- Case IV. Buyer loses the spectrum s_i when he bids truthfully and wins the spectrum s_i when he bids untruthfully. In the former case, he achieves zero utility. Now we have to analyze his utility in the later case. For Case IV to happen, $v_{j,i}^b t_j$ has to be the minimum in group g_i, and $b'_{j,i} > v_{j,i}$. When buyer j increases his bid, the group bid Φ'_i will be higher than Φ_i, and group g_i can win the spectrum s_i. Also, $b'_{j,i} t_j$ will no longer be the minimum, otherwise, buyer j will still lose the spectrum s_i. Let buyer j be the new minimum bid, and we have $v_{j,i}^b t_j < b_{k,i} t_k < b'_{j,i} t_j$, in which buyer k is the new buyer with the minimum bid. Buyer j's utility becomes

$$v_{j,i}^b t_j - b_{k,i} t_k \leq 0$$

Buyer j achieves negative utility when being untruthful. Therefore, he has no incentive to lie.

In summary, buyer j has no incentive to bid untruthfully as his utility cannot be improved.

Proposition 13 *LOTUS is truthful for the buyers.*

Proof It can be easily proved by combining the inter-spectrum truthfulness of Lemma 12 and the intra-spectrum truthfulness of Lemma 15.

Now we consider the truthfulness on the sellers' side. Similarly, we first consider the following two lemmas.

Lemma 16 *If seller i wins, he gets paid the same price regardless of his ask r_i.*

Proof If seller i wins, the received payment is always determined by ϕ_i and the maximum number of desired time slots in g_i, both of which are not affected by r_i.

Lemma 17 *If seller i wins by r_i, it will also win by asking $r'_i < r_i$.*

Proof If seller i asks for a lower price, $r'_i < r_i < \phi_i$. Therefore, the seller will still be able to sell his spectrum to group g_i.

Theorem 3 *LOTUS is truthful for the sellers.*

Proof Let's consider seller i and the buyer group g_i. Assume that seller i asks truthfully as v_i^s, and untruthfully as r_i'. We again consider the four possible cases in Table 2.1.

- Case I. Seller i loses when he bids truthfully and untruthfully, and his utilities are both zero. Therefore, the seller does not gain higher utility by being untruthful.
- Case II. Seller i wins when he bids truthfully and loses when he bids untruthfully. In the former case, he achieves non-negative utility due to individual rationality. In the later case, he achieves zero utility. Therefore, the seller has higher utility when being truthful.
- Case III. Seller i wins when he bids truthfully and untruthfully, and he is paid the same price according to Lemma 16. Since the seller also loses the same valuation for leasing spectrum s_i, his utilities are the same when he bids truthfully and untruthfully.
- Case IV. Seller i loses when he bids truthfully and wins when he bids untruthfully. In the former case, he achieves zero utility. Now we have to analyze his utility in the later case. For Case IV to happen, $v_i^s > \Phi_i$ must be true, and seller i must lower his bid so that $r_i' < \Phi_i$. Seller i's utility becomes $\Phi_i - v_i^s < 0$. Seller i achieves negative utility when being untruthful. Therefore, he has no incentive to lie.

In summary, Seller i has no incentive to bid untruthfully as his utility cannot be improved.

Chapter 5
Future Research Directions

While there is still need for better spectrum auction mechanisms that can efficiently increase spectrum utilization and maintain nice economic properties, researchers start to pay attention to other problems in auction, such as collusion. Although the collusion problem has been studied for homogeneous static spectrum auction, it is not sure whether dynamic spectrum auction and spectrum heterogeneity will raise new challenges, and how to deal with such challenges remains an open issue. In this chapter, we will discuss some possible research directions in the future.

5.1 Collusion in Spectrum Auction

Driven by the hope of higher utility, buyers may collude with each other [9, 31, 57]. The feature of collusion is highly dependent on the commodity being auctioned and the auction mechanism. Traditional auction mechanisms such as secondary auction are quite susceptible to collusion. It is of great importance to detect whether collusion is taking place in an auction, and design auction mechanisms that are resistent to collusion. Conventional collusion forms include buyer collusion, seller collusion and auctioneer collusion.

- *Buyer collusion.* The buyers may form collusion coalition, and the buyers in the coalition agree not to bid against each other. Therefore, the auctioned spectrums can be obtained with low prices, and the buyers in the coalition may divide these spectrums after the auction (e.g., through an auction within the collusion coalition). The most common form of collusion is bidding ring collusion. Multiple buyers form a bidding ring by not to outbidding one another. This will drag down prices and hurt the sellers' revenue. The sellers can avoid a bidding ring collusion by naming an optimal reserve price for their spectrums [33].
- *Seller collusion.* The sellers may bid on their own spectrums, or hire some fake buyers to bid on the spectrums, to increase the price that the buyers have to pay.
- *Auctioneer collusion.* The auctioneer may collude with sellers by overselling the auctioned spectrums. The auctioneer may also collude with buyers by undersell the spectrums.

© The Author(s) 2015

Y. Chen, Q. Zhang, *Dynamic Spectrum Auction in Wireless Communication*,
SpringerBriefs in Electrical and Computer Engineering, DOI 10.1007/978-3-319-14030-8_5

Spectrum auction has brought about new collusion forms due to spectrum reusabil-ity [68]. Assume that we have an interference graph as shown in Fig. 5.1, and there is one spectrum to be sold. The auctioneer groups the buyers as $\{A, D, F\}$, $\{B, E\}$ and $\{C, G\}$. The group $\{A, D, F\}$ is the winner and the other two groups are the losers.

- *Losing group collusion.* If B and E know that they will be grouped together, they can collude to raise their bids higher than their true valuation. In this way, they may outbid group $\{A, D, F\}$ to be winners. As long as the price charged by the auctioneer is lower than their true valuation, they will get positive utility.
- *Sublease collusion.* After the auction, group $\{A, D, F\}$ may decide not to use the spectrum but negotiate with group $\{B, E\}$ to lease the spectrum to them on a unanimous price. In this way, group $\{A, D, F\}$ may reap some profit.
- *Interest group collusion.* Buyers of the same interest group may misreport their interference relationship in order to keep off other buyers. For example, if A, D and G are of the same interest group, A and D may claim that they interfere with F. In this way, the auctioneer may instead group A, D and G together (we assume that the auctioneer is unaware of the true interference relationship), and make them winners.

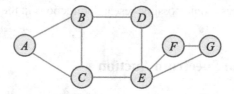

Fig. 5.1. Illustration of the interference graph for collusion

There have been several works on collusion-resistent auction mechanism design [67, 68]. However, they focus on static spectrum auction and treat spectrums as homogeneous. Whether dynamic spectrum auction and spectrum heterogeneity will pose new collusion problems and how to solve them are worth exploring.

5.2 Simultaneous Multiple Round Auction

In this section, we introduce Simultaneous Multiple Round Auction (SMRA) [12, 27, 36], which is different from both the static auction and dynamic auction we introduced before. SMRA is different from the static auction as SMRA has mul-tiple rounds. SMRA is different from the dynamic auction as in SMRA, there will be only one auction result at the end of all rounds, while in dynamic auction, there will be one auction result at each time stage. SMRA is the current practice for FCC to auction long-term spectrum licenses to service providers. In each round, all buyers simultaneously submit their sealed bids. After the round is concluded, the interme-diate bidding results will be announced, including the new bids and their bidders, as well as the highest bid and its bidder. The auction will come to an end if there is no new bid on a spectrum. Due to its complexity, the auction may last from a single day to several months. SMRA is a combinatorial spectrum auction, in which the buyers can bid for combinations of individual spectrums. In a combinatorial auction, the buyer may have higher or lower valuation for the combination than the sum of its parts. For example, a buyer has valuation 1 for spectrum a, and valuation 2 for spectrum b. If spectrums a and b are regarded by the buyer as *substitutes*, he will have lower valuation for the combination $\{a, b\}$, say 2.5, which is smaller than $1 + 2 = 3$. If spectrums a and b are regarded by the buyer as *complements*, he will have higher valuation for the combination $\{a, b\}$, say 3.5, which is higher than 3. In Sect. 3.4.4, we have discussed that adjacent spectrums will bring a buyer higher utility, thus are more likely to be complements. The previously introduced auction mechanisms which allow buyers to bid for multiple spectrums are a special case of combinatorial auction, in which a buyer's valuation for the combination of individual spectrums is exactly the sum of its parts. Apart from susceptibility to collusion, SMRA faces other two challenges.

5.2.1   Optimal Spectrum Allocation

Given the final-round bids of all buyers, the problem of spectrum allocation is: how to find an optimal spectrum allocation that maximizes the revenue of the seller or the auctioneer. For example, we have the bids of three buyers in Table 5.1, without considering spectrum reusability, the optimal spectrum allocation is to assign $\{a, b\}$ to buyer C. When the number of buyers and spectrums increases, the complexity of computing optimal allocation becomes extremely high. In fact, the allocation problem in traditional combinatorial auction is NP-hard. If the spectrum reusability is further considered, the problem will be even more complicated. This makes SMRA, or rather combinatorial auctions, difficult to be applied to dynamic spectrum access, in which the spectrum availability changes rapidly, and therefore fast clearance is required.

Table 5.1 Bids in the SMRA

Buyer/Spectrum	a	b	$\{a,b\}$
A	1	2	4
B	1	3	5
C	2	2	6

5.2.2 Free Riding Problem

In the SMRA, when the "small bidders" want to outbid a "large bidder", free riding problem may arise. For example, we have three buyers and two spectrums. The valuation of each buyer for each spectrum and their combination is shown in Table 5.2. In the first round, buyer A and B both bid 1; buyer C bids 4 and is the provisional winner. In the following rounds, a possible case of ideal bidding is shown in Table 5.3, and buyers A and B are the winners. However, both A and B are selfish, and they try to win their spectrums with as less money as possible. After the results in the first round is revealed, A or B both would prefer a smaller increase, hoping the other will make up for the difference. For instance, if A adopts free riding strategy, the auction results may be the case in Table 5.4. How to avoid such free riding problem and solicit the true valuation of the buyers requires more sophisticated auction design.

Table 5.2 Buyers' valuation for spectrums and their combination

	a	b	$\{a,b\}$
A	8	0	0
B	0	8	0
C	0	0	10

Table 5.3 Ideal bidding process

Round	A bids for a	B bids for b	C bids for $\{a,b\}$
1	1	1	4
2	3	3	8
3	6	6	10

Table 5.4 Free riding bidding process

Round	A bids for a	B bids for b	C bids for $\{a,b\}$
1	1	1	4
2	2	4	8
3	3	8	10

References

1. Spectrum bridge. http://spectrumbridge.com/Home.aspx. Accessed 4 Jan 2015.
2. German spectrum auction ends but prices low. 2010. http://www.rethink-wireless.com/2010/05/21/german-spectrum-auction-ends-prices-low.htm. Accessed 4 Jan 2015.
3. Incentive auctions. 2012. http://www.fcc.gov/topic/incentive-auctions. Accessed 4 Jan 2015.
4. Abhayawardhana, V. S., I. J. Wassell, D. Crosby, M. P. Sellars, and M. G. Brown. 2005. Comparison of empirical propagation path loss models for fixed wireless access systems. *IEEE Vehicular Technology Conference (VTC)* 1:73–77.
5. Al-Ayyoub, Mahmoud, and Himanshu Gupta. 2011. Truthful spectrum auctions with approximate revenue. *IEEE International Conference on Computer Communications (INFOCOM)*, 2813–2821.
6. Alon, Noga, Laszlo Babai, and Alon Itai. 1986. A fast and simple randomized parallel algorithm for the maximal independent set problem. *Journal of algorithms* 7 (4): 567–583.
7. Babaioff, Moshe, and Noam Nisan. 2001. Concurrent auctions across the supply chain. *ACM conference on Electronic Commerce,* 1–10.
8. Babaioff, Moshe, and William E. Walsh. 2005 Incentive-compatible, budget-balanced, yet highly efficient auctions for supply chain formation. *Decision Support Systems* 39 (1): 123–149.
9. Bajari, Patrick, and Jungwon Yeo. 2009. Auction design and tacit collusion in fcc spectrum auctions. *Information Economics and Policy* 21 (2): 90–100.
10. Balakrishnan, Rangaswami, and K. Ranganathan. 2012. *A textbook of graph theory.* Heidelberg: Springer.
11. Brar, Gurashish, Douglas M. Blough and Paolo Santi. 2006. Computationally efficient scheduling with the physical interference model for throughput improvement in wireless mesh networks. *The ACM International International Conference on Mobile Computing and Networking (MobiCom)*, 2–13.
12. Brunner, Christoph, Jacob K. Goeree, Charles A. Holt, and John O. Ledyard. 2006. Combinatorial auctioneering.
13. Chen, Lin, Stefano Iellamo, Marceau Coupechoux, and Philippe Godlewski. 2010. An auction framework for spectrum allocation with interference constraint in cognitive radio networks. *IEEE International Conference on Computer Communications (INFOCOM)*, 1–9.
14. Chen, Yanjiao, Peng Lin, and Qian Zhang. 2014. Lotus: Location-aware online truthful double auction for dynamic spectrum access. *IEEE International Symposium on Dynamic Spectrum Access Networks (DySPAN)*, 510–518.
15. Chen, Yanjiao, Jin Zhang, Kaishun Wu, and Qian Zhang. Tames: A truthful double auction for multi-demand heterogeneous spectrums. *IEEE Transactions on Parallel and Distributed Systems (TPDS)*.
16. Clarke, Edward H. 1971. Multipart pricing of public goods. *Public choice* 11 (1): 17–33.
17. Cramton, Peter. 2002. Spectrum auctions.

© The Author(s) 2015

55

Y. Chen, Q. Zhang, *Dynamic Spectrum Auction in Wireless Communication,*
SpringerBriefs in Electrical and Computer Engineering, DOI 10.1007/978-3-319-14030-8

18. Deshmukh, Kaustubh, Andrew V. Goldberg, Jason D. Hartline, and Anna R. Karlin. 2002. Truthful and competitive double auctions. *AlgorithmsXESA 2002* 361–373. Springer.

19. Edmonds, Jack. 1965. Paths, trees, and flowers. *Canadian Journal of mathematics* 17 (3): 449–467.

20. Erceg, Vinko, Larry J. Greenstein, Sony Y. Tjandra, Seth R. Parkoff, Ajay Gupta Boris Kulic, Arthur A. Julius, and Renee Bianchi. 1999. An empirically based path loss model for wireless channels in suburban environments. *IEEE Journal on Selected Areas in Communications (JSAC)* 17 (7): 1205–1211.

21. Feng, Xiaojun, Yanjiao Chen, Jin Zhang, Qian Zhang, and Bo Li. 2012. Tahes: A truthful double auction mechanism for heterogeneous spectrums. *IEEE Transactions on Wireless Communications (TWC)* 11 (11): 4038–4047.

22. Feng, Xiaojun, Jin Zhang, and Qian Zhang. 2011. Database-assisted multi-ap network on tv white spaces: Architecture, spectrum allocation and ap discovery. In *IEEE Symposium on New Frontiers in Dynamic Spectrum Access Networks (DySPAN)*, 265–276.

23. Feng, Xiaojun, Jin Zhang, and Qian Zhang. 2011. Database-assisted multi-ap network on tv white spaces: System architecture, spectrum allocation and ap discovery. *IEEE International Symposium on Dynamic Spectrum Access Networks (DySPAN)*.

24. Gandhi, Sorabh, Chiranjeeb Buragohain, Lili Cao, Haitao Zheng, and Subhash Suri. 2007. A general framework for wireless spectrum auctions. *IEEE Symposium on New Frontiers in Dynamic Spectrum Access Networks (DySPAN)*, 22–33.

25. Gao, Lin, Jianwei Huang, Ying-Ju Chen, and Biying Shou. 2002. Contrauction: An integrated contract and auction design for dynamic spectrum sharing. CISS 2012.

26. Gao, Lin, Youyun Xu, and Xinbing Wang. 2011. Map: Multiauctioneer progressive auction for dynamic spectrum access. *IEEE Transactions on Mobile Computing (TMC)* 10 (8): 1144–1161.

27. Goeree, Jacob K., Charles A. Holt, and John O. Ledyard. 2006. An experimental comparison of the fccs combinatorial and non-combinatorial simultaneous multiple round auctions. *Prepared for the Wireless Communications Bureau of the Federal Communications Commission*.

28. Goldsmith, Andrea. 2005. *Wireless communications*. Cambridge: Cambridge University Press.

29. Groves, Theodore. 1973. Incentives in teams. *Econometrica: Journal of the Econometric Society* 41 (4): 617–631.

30. Guo, Xingang, Sumit Roy, and W. Steven Conner. 2003. Spatial reuse in wireless ad-hoc networks. *IEEE Vehicular Technology Conference (VTC)* 3:1437–1442.

31. Hendricks, Kenneth, and Robert H. Porter. 1989. Collusion in auctions. *Annales d'Economie et de Statistique* 217–230.

32. Huang, Pu, Alan Scheller Wolf, and Katia Sycara. 2002. Design of a multi-unit double auction e-market. *Computational Intelligence* 18 (4): 596–617.

33. Ji, Zhu, and K. J. Ray Liu. 2008. Multi-stage pricing game for collusion-resistant dynamic spectrum allocation. *IEEE Journal on Selected Areas in Communications (JSAC)* 26 (1): 182–191.

34. Jia, Juncheng, Qian Zhang, and Xuemin Shen. 2008. Hc-mac: A hardware-constrained cognitive mac for efficient spectrum management. *IEEE Journal on Selected Areas in Communications (JSAC)* 26 (1): 106–117.

35. Jia, Juncheng, Qian Zhang, Qin Zhang, and Mingyan Liu. 2009. Revenue generation for truthful spectrum auction in dynamic spectrum access. *The ACM International Symposium on Mobile Ad Hoc Networking and Computing (MobiHoc)*, 3–12.

36. Keuter, Alfons, and Lorenz Nett. 1997. Ermes-auction in germany. First simultaneous multiple-round auction in the european telecommunications market. *Telecommunications Policy* 21 (4): 297–307.

37. Kim, Hyoil, Jaehyuk Choi, and Kang G. Shin. 2011. Wi-fi 2.0: Price and quality competitions of duopoly cognitive radio wireless service providers with time-varying spectrum availability. *IEEE International Conference on Computer Communications (INFOCOM)*, 2453–2461.

38. Kim, Tae-Suk, Hyuk Lim, and Jennifer C. Hou. 2006. Improving spatial reuse through tuning transmit power, carrier sense threshold, and data rate in multihop wireless networks. *The ACM International International Conference on Mobile Computing and Networking (MobiCom)*, 366–377. ACM.

39. Klemperer, Paul. 2002. What really matters in auction design. *The Journal of Economic Perspectives* 16 (1): 169–189.
40. Kothari, Anshul , David C. Parkes, and Subhash Suri. 2005. Approximately-strategyproof and tractable multiunit auctions. *Decision Support Systems* 39 (1): 105–121.
41. Krishna, Vijay. 2009. *Auction theory*. Amsterdam: Academic Press.
42. Lehmann, D., L. I. Oćallaghan, and Y. Shoham. 2002. Truth revelation in approximately efficient combinatorial auctions. *Journal of the ACM (JACM)* 49 (5): 577–602.
43. Luby, Michael. 1985. A simple parallel algorithm for the maximal independent set problem. *ACM symposium on Theory of computing* 1–10.
44. Luby, Michael. 1986. A simple parallel algorithm for the maximal independent set problem. *SIAM journal on computing* 15 (4): 1036–1053.
45. McAfee, R. Preston. 1992. A dominant strategy double auction. *Journal of Economic Theory* 56 (2): 434–450.
46. Micali, Silvio, and Vijay V. Vazirani. 1974. An $o(|v| * |e|)$ algorithm for maximum matching of graphs. *Annual Symposium on Foundations of Computer Science,* 91–98.
47. Milgrom, Paul Robert. 2004. *Putting auction theory to work*. Cambridge: Cambridge University Press.
48. Mu'Alem, Ahuva, and Noam Nisan. 2008. Truthful approximation mechanisms for restricted combinatorial auctions. *Games and Economic Behavior* 64 (2): 612–631.
49. Myerson, Roger B., and Mark A. Satterthwaite. 1983. Efficient mechanisms for bilateral trading. *Journal of economic theory* 29 (2): 265–281.
50. Nguyen, Giao T., Randy H. Katz, Brian Noble, and Mahadev Satyanarayanan. 1996. A trace-based approach for modeling wireless channel behavior. *IEEE Conference on Winter Simulation,* 597–604.
51. Executive Office of the President President's Council of Advisors on Science and Technology. Realizing the full potential of government-held spectrum to spur economic growth.
52. Recommendation ITU-R P.1238-1. 1999. *Propagation Data and Prediction Methods for the Planning of Indoor Radiocomm.*
53. Parzy Marcin, and Hanna Bogucka. 2011. Non-identical objects auction for spectrum sharing in tv white spaces-the perspective of service providers as secondary users. *IEEE International Symposium on Dynamic Spectrum Access Networks (DySPAN),* 389–398.
54. Reardon, Margret. 2010. Fcc unveils national broadband plan. http://news.cnet.com/8301-30686_3-20000453-266.html. Accessed 4 Jan 2015.
55. Reardon, Mary. 2010. Rethinking the wireless spectrum crisis. http://news.cnet.com/8301-30686_3-20005831-266.html. Accessed 4 Jan 2015.
56. Reis, Charles, Ratul Mahajan, Maya Rodrig, David Wetherall, and John Zahorjan. 2006. Measurement-based models of delivery and interference in static wireless networks. *ACM SIGCOMM Computer Communication Review* 36 (4): 51–62.
57. Robinson, Marc S. 1985. Collusion and the choice of auction. *The RAND Journal of Economics* Spring:141–145.
58. Sakai, Shuichi, Mitsunori Togasaki, and Koichi Yamazaki. 2003. A note on greedy algorithms for the maximum weighted independent set problem. *Discrete Applied Mathematics* 126:313–322.
59. Sterbenz, James P. G., Rajesh Krishnan, Regina Rosales Hain, Alden W. Jackson, David Levin, Ram Ramanathan, and John Zao. 2002. Survivable mobile wireless networks: Issues, challenges, and research directions. *ACM workshop on Wireless security,* 31–40.
60. Subramanian, Anand Prabhu, and Himanshu Gupta. 2007. Fast spectrum allocation in coordinated dynamic spectrum access based cellular networks. *IEEE International Symposium on Dynamic Spectrum Access Networks (DySPAN).*
61. Tse, David, and Pramod Viswanath. 2005. *Fundamentals of wireless communication*. Cambridge University Press.
62. Vickrey, William. 1961. Counterspeculation, auctions, and competitive sealed tenders. *Journal of finance* 16 (1): 8–37.

63. Wang, ShiGuang, Ping Xu, XiaoHua Xu, ShaoJie Tang, XiangYang Li, and Xin Liu. 2010. Toda: Truthful online double auction for spectrum allocation in wireless networks. *IEEE International Symposium on Dynamic Spectrum Access Networks (DySPAN)*, 1–10.

64. Wang, Wei, Baochun Li, and Ben Liang. 2011. District: Embracing local markets in truthful spectrum double auctions. *IEEE Communications Society Conference on Sensor, Mesh and Ad Hoc Communications and Networks (SECON)*, 521–529.

65. Wu, Fan, and Nitin Vaidya. 2011. Small: A strategy-proof mechanism for radio spectrum allocation. *IEEE International Conference on Computer Communications (INFOCOM)*, 81–85.

66. Wu, Yongle, Beibei Wang, K. J. Ray Liu, and T. Charles Clancy. 2008. Collusion-resistant multi-winner spectrum auction for cognitive radio networks. *IEEE Global Telecommunications Conference (GLOBECOM)*, 1–5.

67. Wu, Yongle, Beibei Wang, K. J. Ray Liu, and T. Charles Clancy. 2008. A multi-winner cognitive spectrum auction framework with collusion-resistant mechanisms. *IEEE International Symposium on Dynamic Spectrum Access Networks (DySPAN)*, 1–9.

68. Wu, Yongle, Beibei Wang, K. J. Ray Liu, and T. Charles Clancy. 2009. A scalable collusion-resistant multi-winner cognitive spectrum auction game. *IEEE Transactions on Communications* 57 (12): 3805–3816.

69. Yang, Dejun, Xi Fang, and Guoliang Xue. 2011. Truthful auction for cooperative communications. *The ACM International Symposium on Mobile Ad Hoc Networking and Computing (MobiHoc)*, 9.

70. Yang, Dejun, Xi Fang, and Guoliang Xue. 2011. Truthful auction for cooperative communications. *The ACM International Symposium on Mobile Ad Hoc Networking and Computing (MobiHoc)*.

71. Zhou, Xia, Sorabh Gandhi, Subhash Suri, and Haitao Zheng. 2008. Ebay in the sky: Strategy-proof wireless spectrum auctions. *The ACM International Symposium on Mobile Ad Hoc Networking and Computing (MobiHoc)*, 2–13.

72. Zhou, Xia, and Haitao Zheng. 2009. Trust: A general framework for truthful double spectrum auctions. *IEEE International Conference on Computer Communications (INFOCOM)*, 999–1007.

73. Zhou, Xia, and Heather Zheng. 2009. Trust: A general framework for truthful double spectrum auctions (extended). Technical report, UCSB Technical Report.

74. Zhu, Yuefei, Baochun Li, and Zongpeng Li. 2012. Truthful spectrum auction design for secondary networks. *IEEE International Conference on Computer Communications (INFOCOM)*, 873–881.